STAGE 3

Mentals 5

Alan McSeveny Rachel McSeveny Diane McSeveny-Foster

Pearson Australia
(a division of Pearson Australia Group Pty Ltd)
459–471 Church St, Level 1, Building B, Richmond, Victoria, 3121
PO Box 23360, Melbourne, Victoria 8012
www.pearson.com.au

First published 2024 by Pearson Australia
2028 2027 2026 2025
10 9 8 7 6 5 4 3 2 1

Publishers: Sophie Matta and Kerry Nagle
Project Manager: Michelle Thomas
Production Editor: Laura Rentsch
Editor: Rachel Elliott
Designer: Anne Donald
Proofreader: Ann M. Philpott
Rights & Permissions Editor: Alice McBroom
Cover art: Michael Barter
Illustrator: Michael Barter
Publishing services: Jit-Pin Chong
Printed in Australia by Pegasus Media and Logistics

ISBN 978 0 6557 0912 1
Pearson Australia Group Pty Ltd ABN 40 004 245 943

Acknowledgement of Country
Pearson respects and honours Aboriginal and Torres Strait Islander Elders past, present and future. We acknowledge the stories, traditions and living cultures of the Traditional Custodians of the lands on which our company is located and where we conduct our business. Pearson is committed to honouring Australian Aboriginal and Torres Strait Islander peoples' unique cultural and spiritual relationships to the land, waters and seas and their rich contribution to society.

Aboriginal and Torres Strait Islander peoples are advised that this text may contain images, voices and names of deceased persons.

Introduction

Using the Mentals Books

This book reviews content from the Signpost Student Book. It is used most effectively when it aligns with the suggested program in the Student Book contents.

Each unit of the Mentals Book is programmed to review Student Book content for the previous two weeks (the Suggested Program overview can be found in the Teacher Resource). For example, Unit 15 of the Mentals Book can be set as homework to review weeks 13 and 14 of the Student Book while week 15 is being taught.

Mixed-topic questions

The units present questions in a mixed-topic format to encourage thorough understanding and continuous review.

Graded questions

- Column 1: easier
- Columns 2 and 3: harder
- Column 4: Extension and Challenge

Presentation

- Number facts are reinforced to encourage instant recall.
- Essential skills are explained.
- The Arithmetic card (page 5) is a useful teaching tool for practising basic number skills.
- ID cards (pages 6 to 9) review the mathematical terms students need to learn.
- Measurement benchmarks and tables of number and measurement (pages 84 and 85) are provided so that students can learn important facts and estimate measurements effectively.

Motivation

- There are two lizards hidden on each page for students to find.
- The header allows students to record their score.

Extra activities

- Problem solving **strategies** are introduced in a carefully planned sequence throughout the series.

- Important concepts from **Number and algebra** and **Measurement and geometry** are explored.

- **Measurement** concepts and activities are introduced and investigated.

- **Statistics and probability** concepts (Data and chance) are presented for revision and extension.

- A **tables** program for each of the four operations is included.
- It is important for students to learn addition and multiplication tables by heart.

5 Contents

Unit activities

Unit	Content	Extra Activity
1:1/2 **1:3/4**	× 2, × 6 Personal measurements	× tables Measure
2:1/2 **2:3/4**	× 10, × 5 × 3, × 4	× tables × tables
3:1/2 **3:3/4**	× 3, × 6 × 7	× tables × tables
4:1/2 **4:3/4**	× 8 × 9	× tables × tables
5:1/2 **5:3/4**	÷ 2, ÷ 4 ÷ 3, ÷ 6	÷ tables ÷ tables
6:1/2 **6:3/4**	Place value × 9, ÷ 9	Concept ×, ÷ tables
7:1/2 **7:3/4**	Perimeter Language	Measure ID card C
8:1/2 **8:3/4**	× 10, × 5 Language	× tables ID card A
9:1/2 **9:3/4**	Using a graph Area and perimeter	Concept Strategy time
10:1/2 **10:3/4**	Rounding off The jump strategy	Concept Strategy time
11:1/2 **11:3/4**	Roman numerals × 9	Concept × tables
12:1/2 **12:3/4**	Perimeter ÷ 9 and × 4, × 7, × 9	Measure ÷, × tables
13:1/2 **13:3/4**	Perimeter Language	Measure ID card D
14:1/2 **14:3/4**	Division Rounding off money	÷ tables Concept
15:1/2 **15:3/4**	Language × 2, × 5, × 4, × 10, × 0, × 1	ID card B × tables
16:1/2 **16:3/4**	Language Division with remainders	ID card D Concept
17:1/2 **17:3/4**	× 3, × 6 – 3, – 7	× tables – tables
18:1/2 **18:3/4**	Compass points × 8, × 6	Concept × tables
19:1/2 **19:3/4**	Compass directions 2D space	Concept Concept

Unit	Content	Extra Activity
20:1/2 **20:3/4**	– 6, – 8 Is this game fair?	– tables Chance
21:1/2 **21:3/4**	× 7, × 8 × 9, × 7	× tables × tables
22:1/2 **22:3/4**	Problem solving ÷ 7, ÷ 8	Strategy time ÷ tables
23:1/2 **23:3/4**	Language Problem solving	ID card C Strategy time
24:1/2 **24:3/4**	Dot plot 24-hour time	Data Measure
25:1/2 **25:3/4**	Dot plot × 11	Data × tables
26:1/2 **26:3/4**	Division Factors of a number	÷ tables Concept
27:1/2 **27:3/4**	Division Patterns	÷ tables Concept
28:1/2 **28:3/4**	Factors Mass	Concept Strategy time
29:1/2 **29:3/4**	Language Roman numerals	ID card A Concept
30:1/2 **30:3/4**	Averages × 6, × 7, × 8	Concept × tables
31:1/2 **31:3/4**	× 3, × 5, × 9 Codes	× table Concept
32:1/2 **32:3/4**	Language Fractions (subtraction)	ID card B Concept
33:1/2 **33:3/4**	Magic squares Fractions (+ and –)	Concept Concept
34:1/2 **34:3/4**	Problem solving Comparing chance	Strategy time Chance
35:1/2 **35:3/4**	Survey Travel graph	Concept Concept
36:1/2 **36:3/4**	15 –, 16 – × 4, × 9	– tables × tables
37:1/2 **37:3/4**	Patterns Personal measurements	Strategy time Measure
Answers	These can be found in the middle of this book on pages A1 to A16.	

Arithmetic card

	A	B	C	D	E	F	G	H	I	J	K	L	M
1	20	9	110	20	6	11	$100	$\frac{67}{100}$	0·52	97%	$\frac{56}{100}$	0·68	67
2	16	1	170	50	14	21	$32	$\frac{3}{100}$	0·2	52%	$\frac{8}{100}$	0·7	62
3	12	6	200	10	2	27	$48	$\frac{75}{100}$	0·75	39%	$\frac{27}{100}$	0·8	50
4	18	2	130	70	18	25	$90	$\frac{46}{100}$	0·39	74%	$\frac{15}{100}$	0·4	41
5	13	8	160	100	10	13	$117	$\frac{5}{100}$	0·4	63%	$\frac{88}{100}$	0·75	33
6	15	5	180	30	16	17	$150	$\frac{9}{100}$	0·98	6%	$\frac{42}{100}$	1	26
7	19	3	150	80	4	29	$76	$\frac{14}{100}$	0·67	18%	$\frac{74}{100}$	0·69	19
8	14	7	190	60	12	15	$28	$\frac{83}{100}$	0·5	45%	$\frac{66}{100}$	1·2	14
9	11	4	140	90	20	23	$55	$\frac{6}{100}$	0·13	87%	$\frac{39}{100}$	0·13	7
10	17	10	120	40	8	19	$85	$\frac{1}{100}$	0·8	21%	$\frac{95}{100}$	0·88	3

How to use this card

If students were told to 'subtract B from C', they would write:

1. 110 – 9 = 101
2. 170 – 1 = 169
3. 200 – 6 = 194
4. 130 – 2 = 128
5. 160 – 8 = 152
6. 180 – 5 = 175
7. 150 – 3 = 147
8. 190 – 7 = 183
9. 140 – 4 = 136
10. 120 – 10 = 110

Other instructions might be:

- **What is left from $200 if I spend the amount in column G?**
- **What multiplied by 10 gives column C?**
- **Write column J as a decimal.**
- **Add column B to column F.**
- **Multiply column B by 4.**
- **Add columns A and C.**
- **Add columns H and K.**
- **Subtract column B from column A.**
- **Halve column E.**

The applications of this card are endless.

ID card A

Do not write on this card.

1. **mm** stands for m______.
2. **cm** stands for c______.
3. **m** stands for m______.
4. **km** stands for k______.
5. **mL** stands for m______.
6. **L** stands for l______.
7. **g** stands for g______.
8. **kg** stands for k______.
9. **s** stands for s______.
10. **min** stands for m______.
11. **h** stands for h______.
12. **cm²** stands for s______ c______.
13. **m²** stands for s______ m______.
14. (1m × 1m) 1 square m ______
15. (100 m × 100 m, 1 ha) This area is 1 h______.
16. **ha** stands for h______.
17. **km²** stands for s______ k______.
18. (1 cm × 1 cm × 1 cm) 1 cubic c______.
19. **°C** stands for d______ C______.
20. **5 groups of 7** means 5 ____ 7
21. **4 rows of 10** means 4 ____ 10
22. **Share 20 among 5** means 20 ____ 5
23. **How many groups of 6 in 42?** means 42 ____ 6
24. **2, 4, 6, …** are the e ______ numbers.
25. **1, 3, 5, …** are the o ______ numbers
26. **72 086** has 5 d______.
27. **The factors of 8** are 1, 8, ______ and ______.
28. **7, 14, 21, …** are multiples of ______.
29. **35, 40 and 45** are all divisible by ______.
30. $5\overline{)47}$ = **9 r 2** 'r' means r______.

See page A1 for answers.

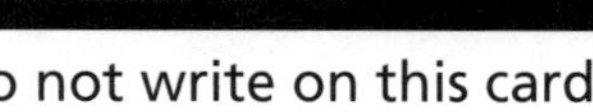

ID card B

Do not write on this card.

1 A ray is part of a l______.

2 p______ lines

3 p______ lines

4 h______ line

5 v______ line

6 c______ or v______ of an angle

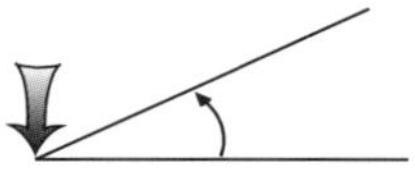

7 a______ of an angle

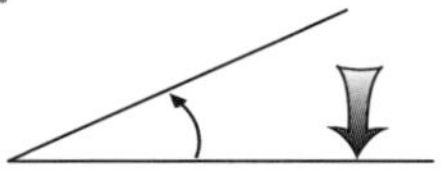

8 a______ angle

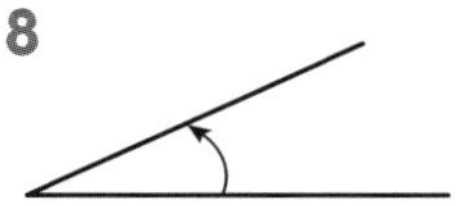

9 r______ angle

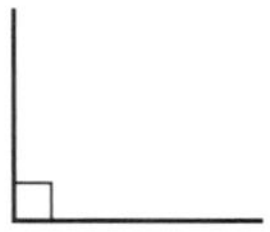

10 o______ angle

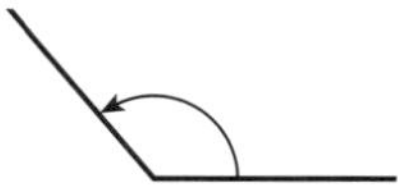

11 s______ angle

12 r______ angle

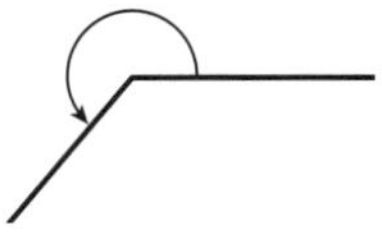

13 r______

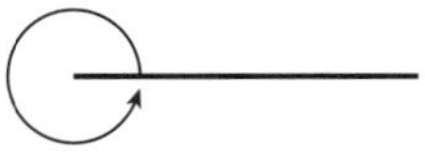

14 t______

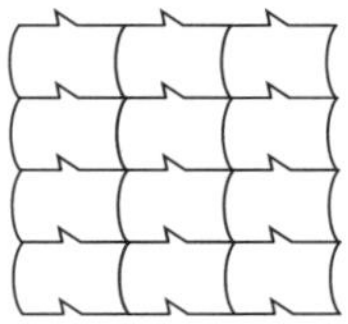

15 f______

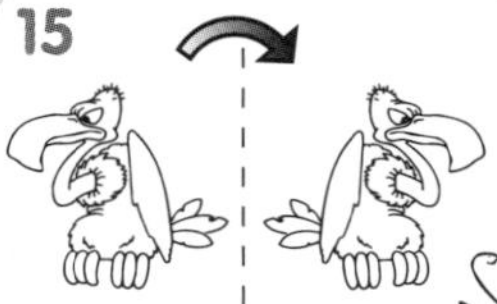

16 s______

17 t______

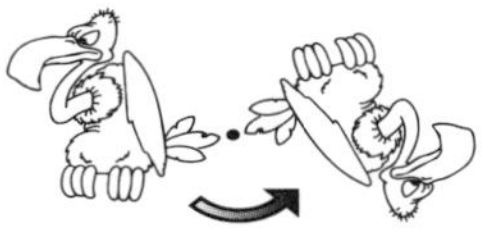

18 axis of s______

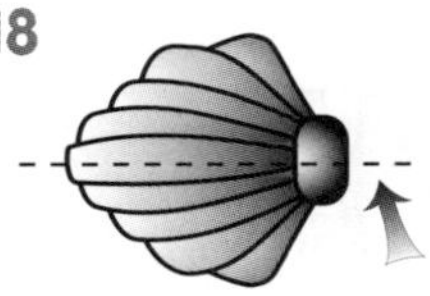

19 a______ of symmetry

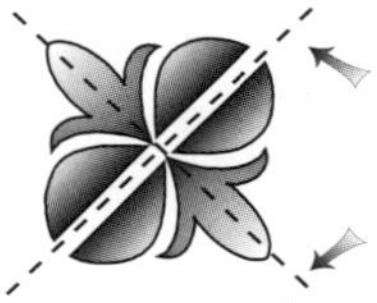

20 0 1 2 3 4

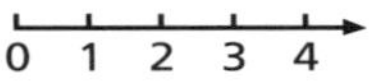

n______ line

21 n______ w______  e______ s______

NW NE SW SE

22 north west ______ ______ ______ ______ ______ ______ ______

N W E S

23 4 3 2 1 0 A B C D E F

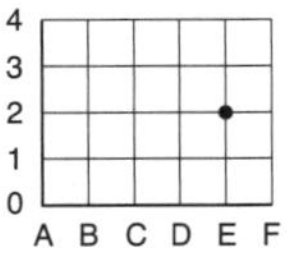

The c______ of the dot are **E2**.

24 Car colour

Red	III
Blue	~~IIII~~ II
Green	~~IIII~~ ~~IIII~~

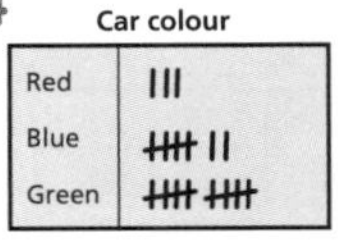

This is a t______.

25 Balls lost

May ◯◯◯
June ◯◯
July ◯◯◯◯

◯ stands for one ball

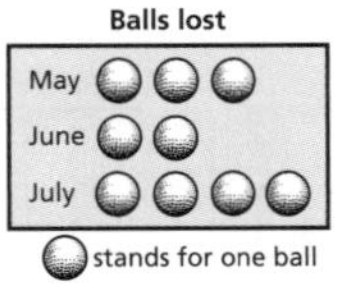

p______ graph

26 Gold stars earned

David, Alana, Luke, Naomi

0 4 8 12

Number

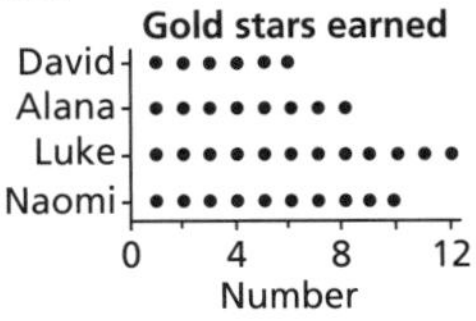

d______ plot

27 Colour of dress

1 2 3 4 5

A B C

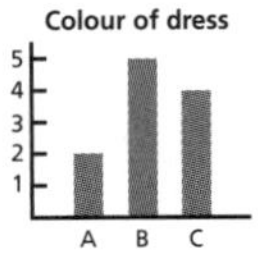

c______ graph

28 Houses sold

16 12 8 4

Mon Tue Wed

l______ graph

29 Colour of balls

blue, green, yellow, red

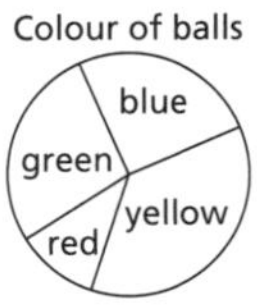

s______ or p______ graph

30 People present

Adults, Girls, Boys

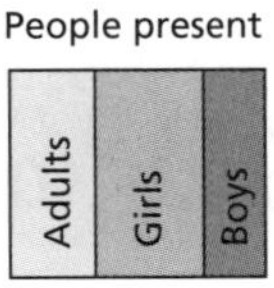

d______ b______ graph

See page A1 for answers.

ID card C

Do not write on this card.

1

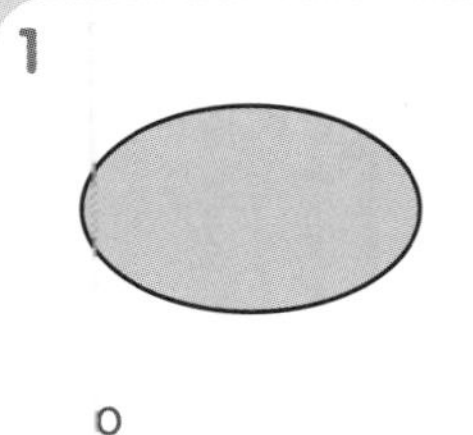

o ________

2

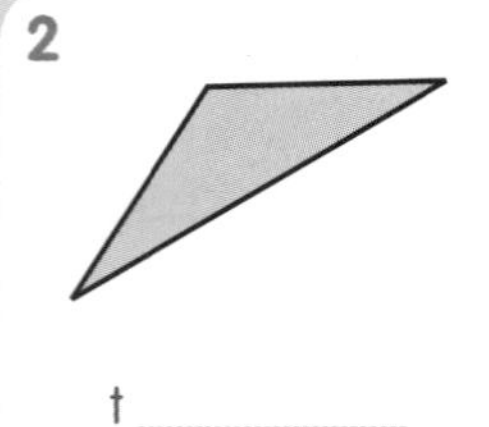

t ________

3

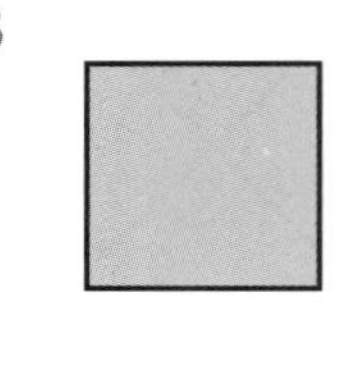

s ________

4

r ________

5

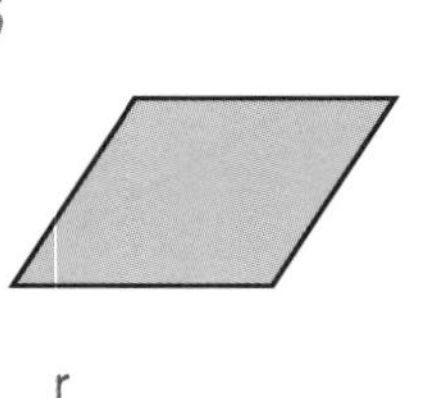

r ________

6

t ________

7

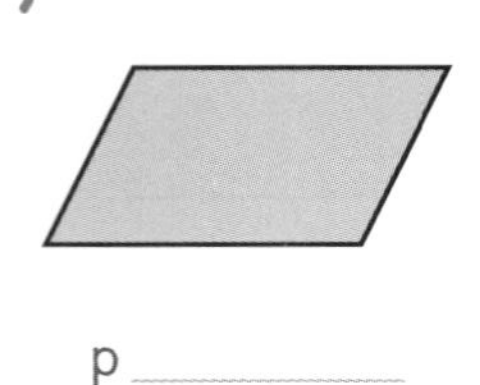

p ________

8

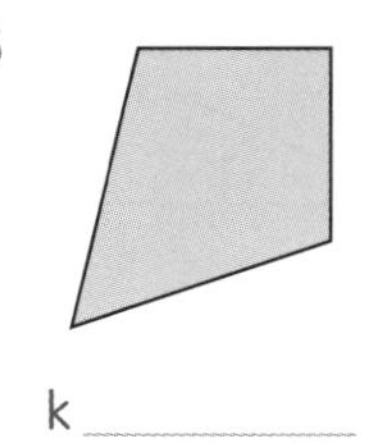

k ________

9

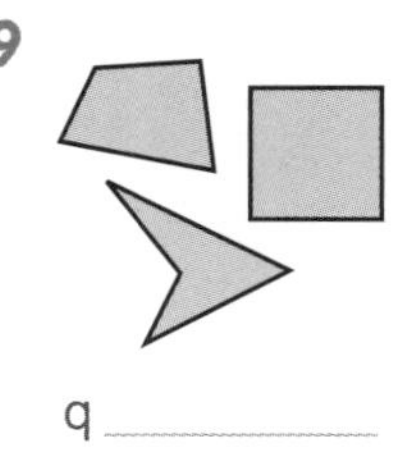

q ________

10

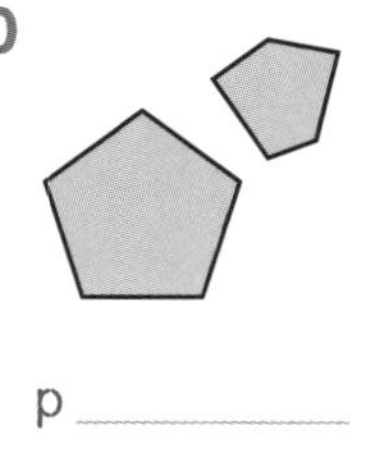

p ________

11

h ________

12

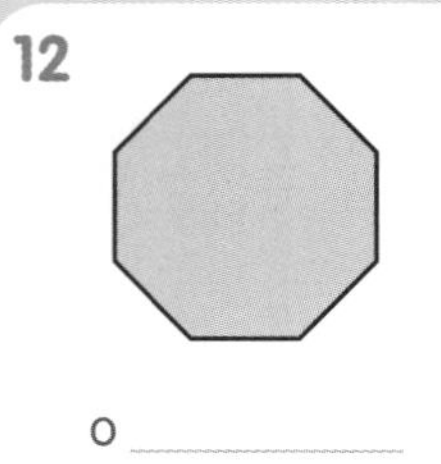

o ________

13

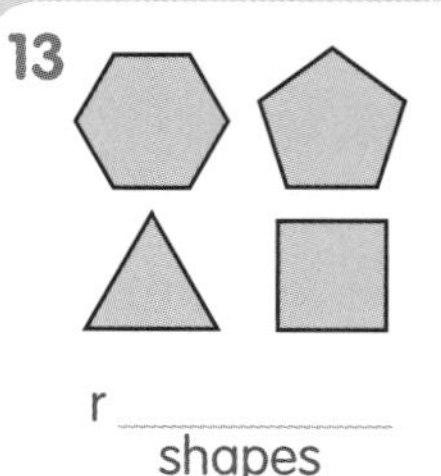

r ________ shapes

14

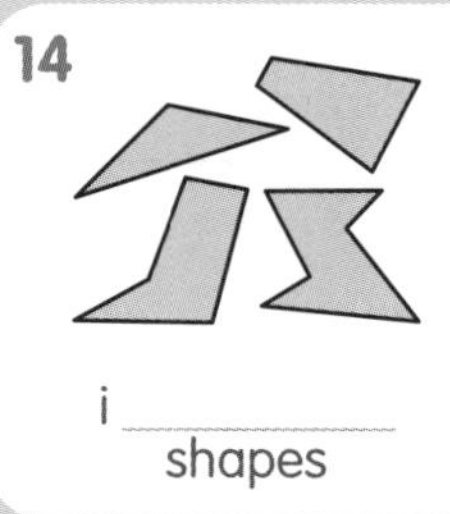

i ________ shapes

15

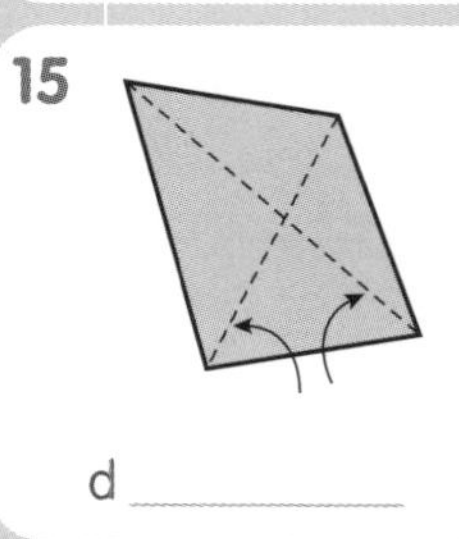

d ________

16

0·5

decimal

p ________

17

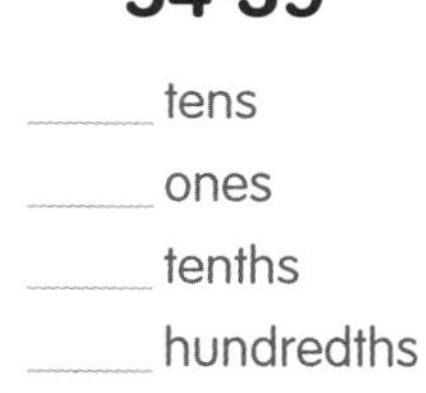

54·39

____ tens

____ ones

____ tenths

____ hundredths

18

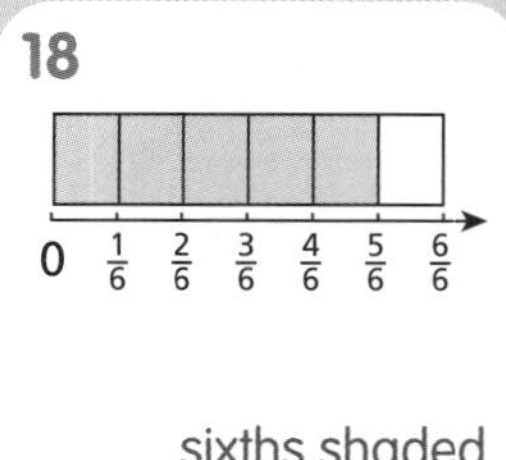

____ sixths shaded

19

$\frac{3}{8}$

means

3 out of ____.

20

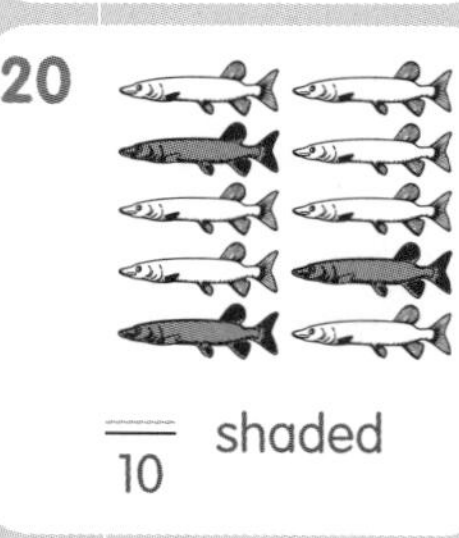

$\frac{\_\_}{10}$ shaded

21

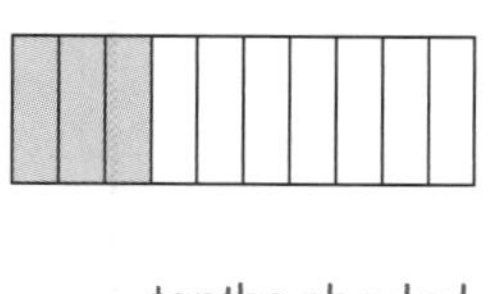

____ tenths shaded

22

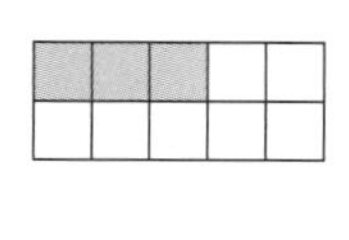

0·____ shaded

23

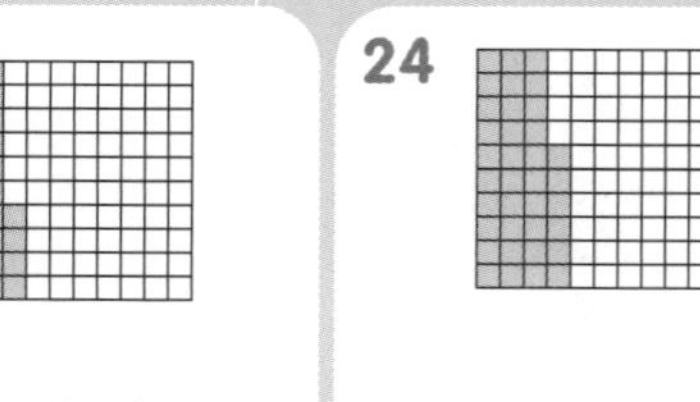

$\frac{\_\_}{100}$ shaded

24

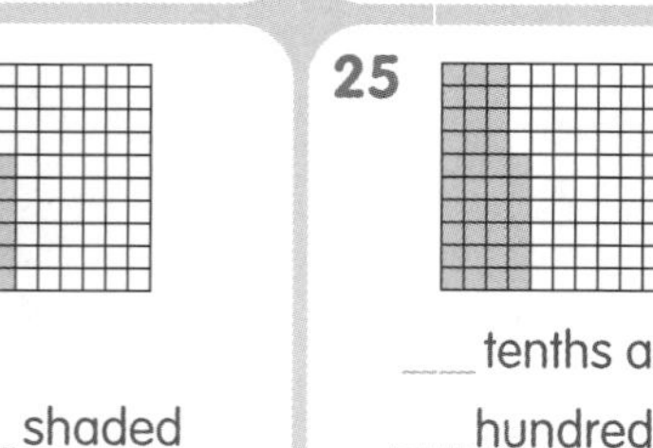

0·____ shaded

25

____ tenths and

____ hundredths

26

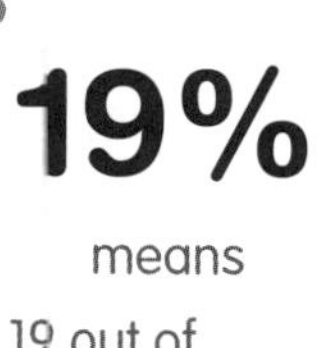

19%

means

19 out of ____.

27

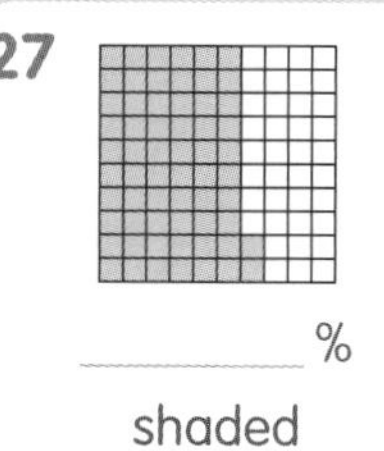

________ %

shaded

28

1 and 36 hundredths

is ____·____

29

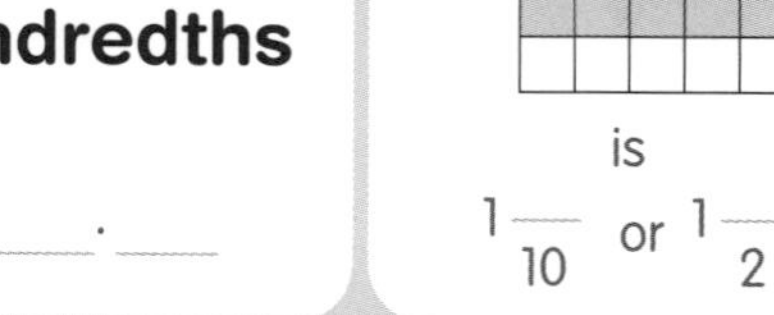

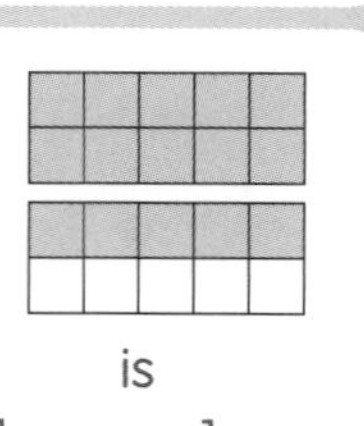

is

$1\frac{\_\_}{10}$ or $1\frac{\_\_}{2}$

30

Complete:

$0{\cdot}3 = \frac{\_\_}{10}$

$0{\cdot}30 = \frac{\_\_}{10}$

See page A1 for answers.

ID card D

Do not write on this card.

1	2	3	4	5
f _____	v _____ or c _____	e _____	c _____	r _____ p _____
6	**7**	**8**	**9**	**10**
t _____ p _____	h _____ p _____	t _____ p _____	s _____ p _____	r _____ p _____
11	**12**	**13**	**14**	**15**
b _____	c _____	c _____	s _____	net of a c _____
16	**17**	**18**	**19**	**20**
net of a s _____ p _____	net of a c _____	net of a c _____	net of a t _____ p _____	Side view is: 1 2 3
21	**22**	**23**	**24**	**25**
Front view is ___. 1 2 3	Top view is ___. 1 2 3	**am** means _____ noon	**pm** means _____ noon	**7:30 pm** in 24-hour time is ___:___
26	**27**	**28**	**29**	**30**
05:40 in digital time is ___:___ am	p _____	t _____	c _____	t _____ m _____

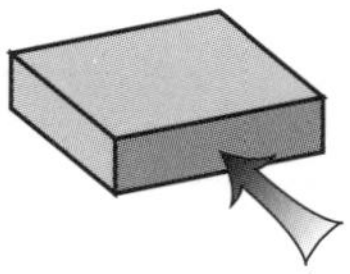

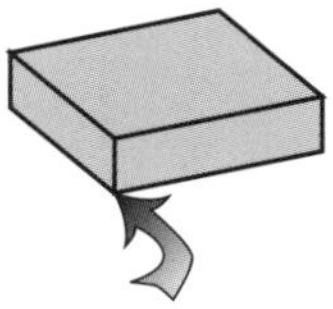

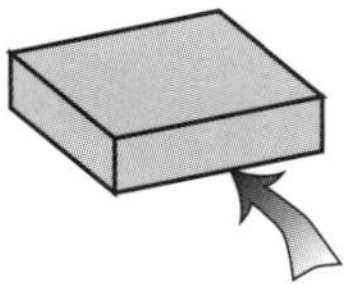

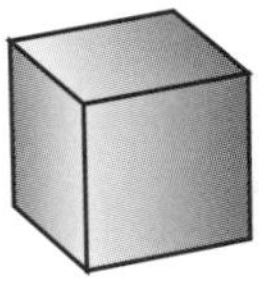

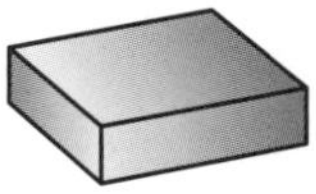

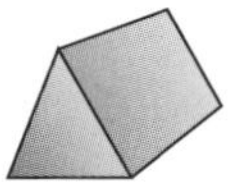

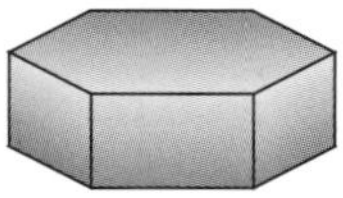

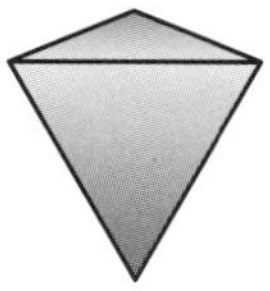

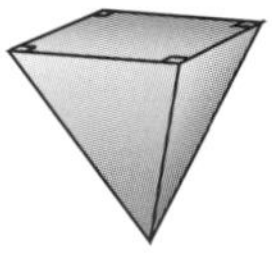

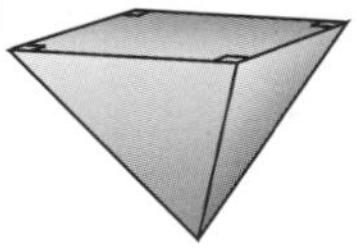

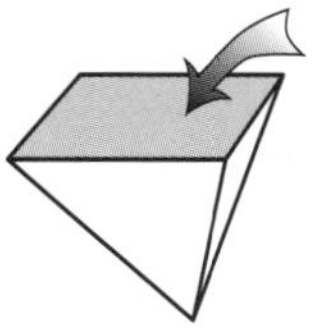

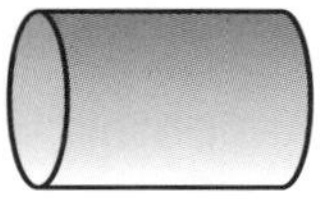

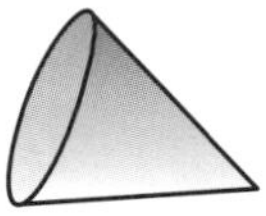

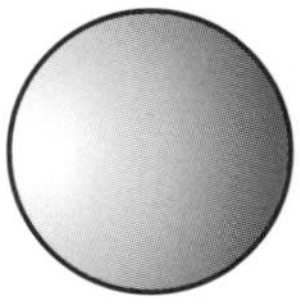

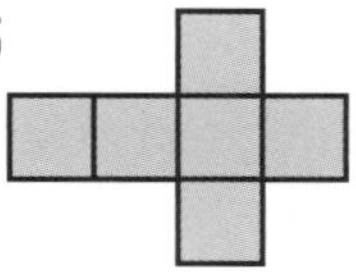

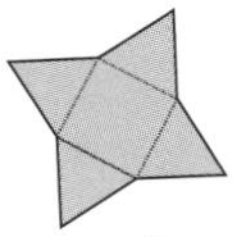

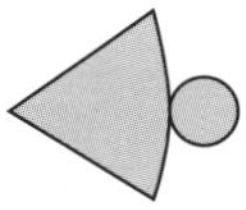

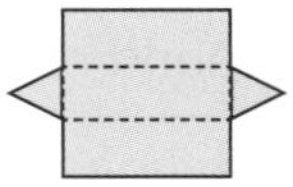

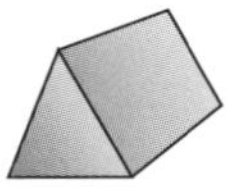

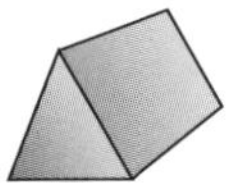

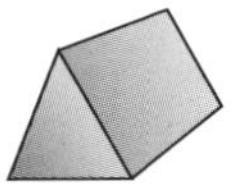

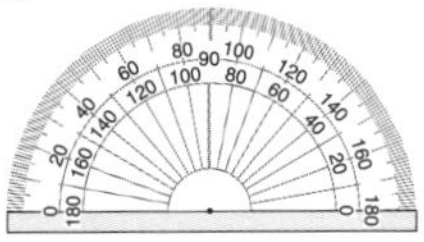

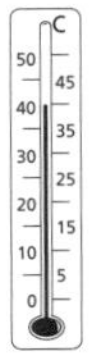

See page A1 for answers.

1:1 out of 19

1. 5 × 10 ______
2. 90 ÷ 9 ______
3. 16 − ______ = 9
4. 7 × ______ = 14
5.
$$\begin{array}{r} 68 \\ -\,45 \\ \hline \end{array}$$
6. 20 divided by 4. ______
7. 5 times 100. ______
8. 51 plus 34. ______
9. Half of 64. ______
10.
$$\begin{array}{r} 462 \\ +\,315 \\ \hline \end{array}$$

11. If 5 × 6 = 30 then:

    a 30 ÷ 5 = ______ **b** 30 ÷ 6 = ______

12. Colour $3\frac{1}{2}$ trapeziums.

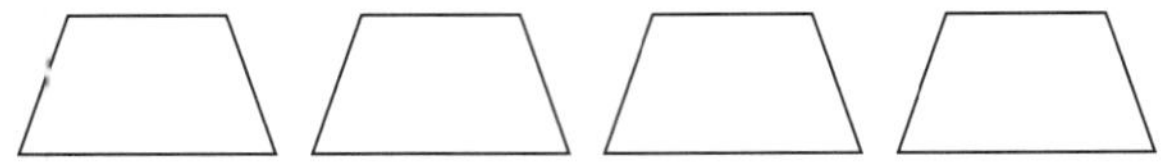

13. Bridge to 10 to find:

    a 27 + 5 = ______ **b** 18 + 7 = ______

14. What is the change from $20 if I spend $3.50? ______
15. The number 3·24 has ______ ones, ______ tenths and ______ hundredths.
16. How many sides on 3 pentagons? ______
17. Multiply each number by 100.

    a 6 ______ **b** 9 ______ **c** 7 ______

18. Of 38 biscuits, 28 were eaten.

    How many were left?

______

19. 75 hundreds + 4 tens + 2 ones = ______

1:2 out of 16

1. 36 × 10 ______
2. 28 + 6 = 30 + ______
3. 8 × ______ = 32
4. 48 ÷ 8 ______
5.
$$\begin{array}{r} 536 \\ +\,389 \\ \hline \end{array}$$
6. 40 shared by 4. ______
7. 40 ÷ 5 + ______ = 15
8. 56 − 23 = 36 − ______
9. Double 3 × 5. ______
10.
$$\begin{array}{r} 475 \\ -\,259 \\ \hline \end{array}$$

11. Explain how you could use this diagram to find 3 × 28 = ______

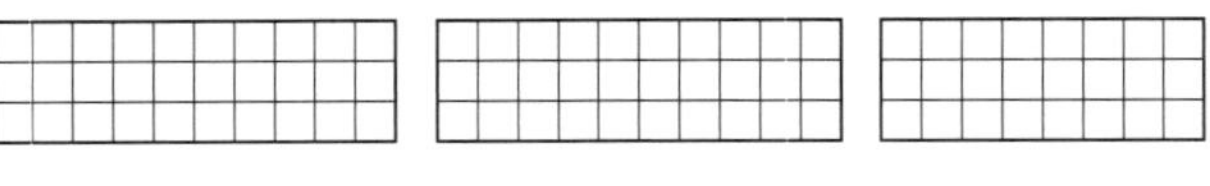

______

______

12. Find the area of this square.

    ______ cm^2

    What is the area of one of the large triangles? ______

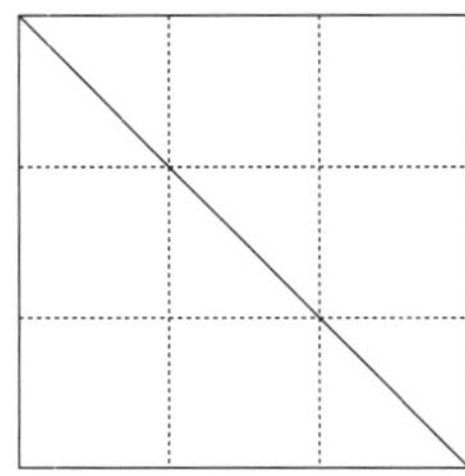

13. **a** How many kilograms in 3000 grams? ______

    b How many kilograms in 5000 grams? ______

14. How could you put 42 people into equal groups?

    ______

    ______

15. What is $\frac{1}{2}$ of 26? ______
16. How many tens could be taken from 647? ______

× tables

How fast are you?

× 2	6	3	8	1	9	7	4	2	10	5

× 6	2	7	4	10	1	7	2	9	5	6

All answers are even.

Even numbers end in 0, 2, 4, 6 or 8.

1:3 — out of 10

1. What fraction is shaded?

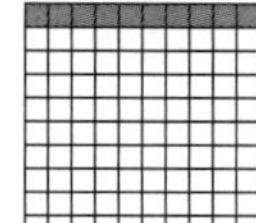

_____ out of _____

2. 6 photos fill a page. How many pages can be filled by 24 photos? _____

3. Which is smaller, 3632 or 3622? _____

4. 4 girls shared 44 grapes. How many grapes was each girl given? _____

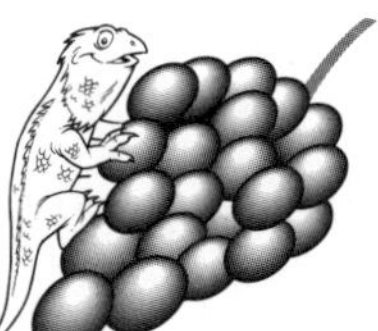

5. Add 20 each time.

   14, _____, _____, _____, _____, _____

6. A farmer had 87 sheep and sold 52. How many did she have left? _____

7. **a** 58 + 36 = _____

   (number line starting at 58)

   b 92 − 48 = _____

   (number line starting at 92)

8. Use the compensation strategy to find:

   a 36 + 97 = _____ **b** 147 + 96 = _____

9. Subtract tens to each side to make this easier.

   78 − 36 = _____ − _____ = _____

10. 3, 6, 9, _____, _____, _____, _____, _____

1:4 — out of 8 — Extension

1. 349 + 34 _____

   (number line starting at 349)

2. **a** 111 more than 1232. _____

   b 212 more than 3400. _____

3. Alex had 4 ice-cream containers with 4000 mL of ice-cream in each. How many litres did she have? _____

4. If 3 boys collected 10 shells each and 5 girls collected 8 shells each, how many shells were collected? _____

5. What is the total length of four pieces of ribbon of lengths 5 m, 2 m 36 cm, 134 cm, and 28 cm? _____

6. If 8 + △ = 13 and 4 × □ = 16, what is △ × □ equal to? _____

7. The smallest number of coins that will total exactly $10.75. _____

8. How many different kinds of outfits can you make from 2 shirts and 5 pants? _____

Challenge

Draw a prism and describe it.

_____

Fill out this table about yourself, a relative or a friend.

Name: _____ **Date:** _____

Age: _____	Mass: _____ kg	Shoe size: _____
Height: _____ cm	Waist: _____ cm	Neck size: _____ cm

2:1 □ out of 19

❶ 7 × 3 ______

❷ 6 × 4 ______

❸ 6 × 8 ______

❹ 3 × 9 ______

❺ 49 − 37

❻ 5 × 3 + 5 ______

❼ Double 5 × 4. ______

❽ 70 shared by 7. ______

❾ 5 × ______ = 25

❿ 639 + 245

⓫ If 6 × 7 = 42, then 7 × 6 = ______.

⓬ **a** If 3 × 4 = ______, then 4 × 4 = ______.
b Does 4 × 4 equal 3 × 4 + 4? ______

⓭ Colour 7 tenths of this hundreds block and write the decimal.

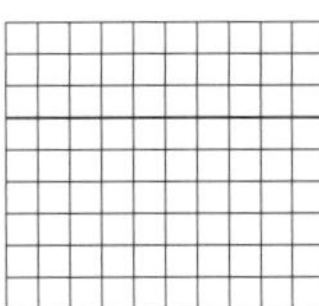

______ · ______

⓮ Is 0·9 larger than 0·09? ______

⓯ **a** ______ × 5 = 35 **b** ______ × 3 = 18
c 35 ÷ 5 = ______ **d** 18 ÷ 3 = ______

⓰ 56 hundreds + 46 ones = ______

⓱ Write the short form for 456 millilitres. ______

⓲ Count on from $5.85 to find the change given for $10. ______

+ □ + □ + □ + □

$5.85 → ______ → ______ → ______ → ______

⓳ Colour $\frac{1}{4}$ of this rectangle.

2:2 □ out of 15

❶ 7 × 1000 ______

❷ 53 × 10 ______

❸ 400 ÷ 10 ______

❹ 6 × 30 ______

❺ 700 − 475

❻ 64 shared by 8. ______

❼ 5 × ______ = 45

❽ 4 × ______ = 28

❾ $30 + 50c + 10c ______

❿ 382 + 617

⓫ Explain how you could use this diagram to find

13 × 3 = ______

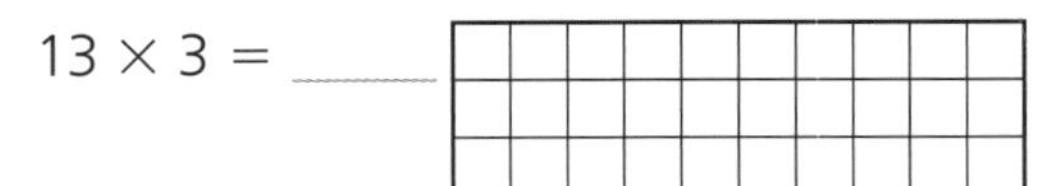

______

______

⓬ **a** 527 + 96 = 527 + 100 − ______ = ______
b 374 + 98 = ______ + 100 − ______ = ______
c 734 + 97 = ______ + 100 − ______ = ______

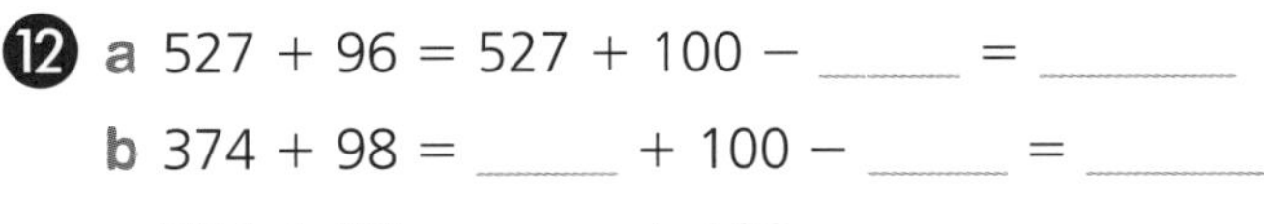

⓭ 10 pet bunnies cost $800.
How much would each pet bunny cost? ______

⓮ Colour $2\frac{1}{4}$ of these shapes.

⓯ Write these numbers on the place value chart.

a 57 342 **b** 78 546 **c** 32 254

	Thousands	Hundreds	Tens	Units
a				
b				
c				

× tables

× 10: 5, 3, 10, 7, 4, 8, 2, 9, 6, 1

× 5: 5, 3, 10, 7, 4, 8, 2, 9, 6, 1

8 × 5 is half of 8 × 10.

2:3 ☐ out of 10

1. 57 thousands + 35 tens + 9 = ______
2. Write the numeral for nine hundred and fifty-six thousand, nine hundred and forty-one. ______
3. Each Monday for 4 weeks, Lien picked 6 flowers.

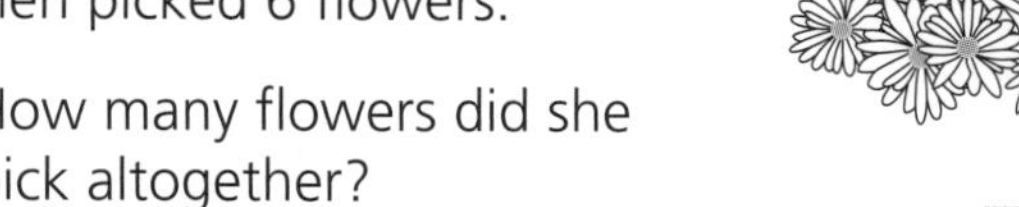

   How many flowers did she pick altogether? ______
4. **a** 21 days = ______ weeks

   b 70 days = ______ weeks
5. Shade one third of the snakes.

6. How many hours from:

   a 4 am to 7 am? ______

   b 12 noon to 3 pm? ______
7. In this hundred square:

   a part shaded is ______ hundredths.

   b part not shaded is ______ out of ______.
8. \$0.40 + \$0.15 + \$0.20 ______
9. Sophie made 34 Anzac biscuits and Freida made 28. How many biscuits altogether? ______
10. 36 thousands + 47 tens = ______

2:4 Extension ☐ out of 10

1. **a** 25 + 25 + 25 + 25 ______

   b 25 × 3 ______

   c 25 × 5 ______

2. How many legs on 3 cats, 8 birds and 5 snakes? ______
3. After losing \$18 and finding only \$12, I had \$16 left. How much did I start with? ______
4. What must we add to \$5.75 to make \$10.50? ______
5. In two dozen eggs, there were 10 more duck eggs than chicken eggs. How many duck eggs? ______
6. **a** How many sides on 11 octagons? ______

   b (7 × 8) + (7 × 8) ______
7. 6 × 10 × 9 ______
8. 100 − 35 − 35 ______
9. 50 − 15 − 15 − 15 ______

10. If # means 'add 3' and ~ means 'subtract 2', then 8 # # # ~ ~ ~ = ______

Challenge

Write what you know about the number 534·6.

______

______

______

______

______

______

 • *AUSTRALIAN SIGNPOST MATHS NSW 5 MENTALS* • ISBN 978 0 6557 0912 1

3:1 ☐ out of 17

1. 15 + 6 ______
2. 45 + 6 ______
3. 18 ÷ 2 ______
4. 8 × 100 ______
5. $\begin{array}{r} 535 \\ +\ 257 \\ \hline \end{array}$
6. 23 minus 9. ______
7. 24 divided by 3. ______
8. Take 10 from 57. ______
9. 8 groups of 5. ______
10. $\begin{array}{r} 73 \\ -\ 48 \\ \hline \end{array}$

11. 

    Write the total value of these coins in decimal form. ______
12. 7 rows of 4. ______
13. Does a spoon weigh more than 1 kilogram? ______
14. How many minutes before 5 o'clock?

    a  ______

    b  ______
15. Give the unit of measure for the length:

    a of a table 1·5 ______

    b of a mobile phone 10·5 ______
16. a 36 + 39 = ____ + 40 = ______

    b 27 + 47 = ____ + 50 = ______
17. 57 hundreds + 5 tens + 35 hundredths = ______

3:2 ☐ out of 19

1. 7 + ______ = 20
2. 7 × ______ = 56
3. 3 × ______ = 24
4. 36 ÷ 9 = ______
5. $\begin{array}{r} 61 \\ -\ 46 \\ \hline \end{array}$
6. 25 divided by 5 = ______
7. 13 + 7 = 10 + ______
8. 46 − 29 = ______ − 9
9. 89 − 53 = ______ − 3
10. $\begin{array}{r} 683 \\ +\ 493 \\ \hline \end{array}$

11. Minutes in 2 hours. ______
12. Days in 6 weeks. ______
13. I saw 84 birds. 26 flew away. How many can I see now? ______
14. John had 32 coins. 14 of these were Australian. How many were not Australian coins? ______
15. Is this a net of a cube? ______

    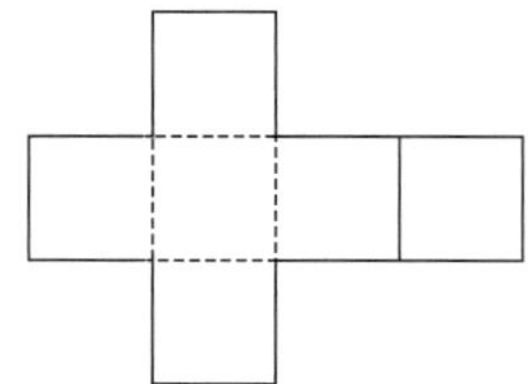
16. Is the area of a door more or less than one square metre? ______
17. On a cube, how many:

    a faces? ______ b edges? ______
18. Complete this pattern if the dotted lines are lines of symmetry.

    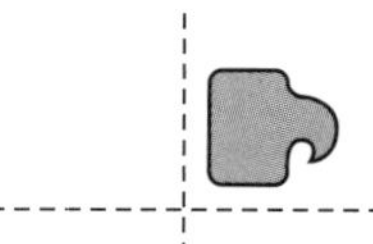
19. What am I if my 4 faces are all triangles? ______

× tables

2	6	9	1
10	× 3		4
8	5	3	7

9	6	3	7
1	× 6		8
10	2	5	4

I know my tables. Do you?

3:3 ☐ out of 12

1. $8\overline{)48}$ 2. $8\overline{)64}$ 3. $8\overline{)24}$

4. 'Consective' means following on one after the other. Name 2 consecutive months that have the same number of days.

____________ ____________

5. The change from $50 if I buy a watch for $37.

____

6. Could this shape be folded to make a cube?

____

7. Greg had ten 20c coins. How much money did he have? ____________

8. Write 478 391 in words. ____________

9. Label each angle with its name.

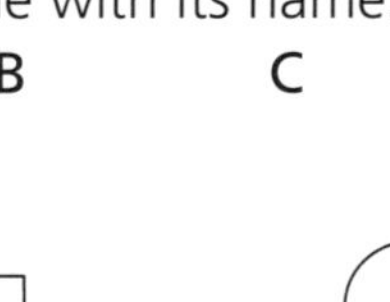
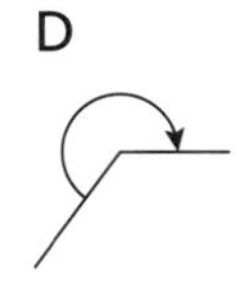

A ______ B ______ C ______ D ______

10. **a** The value of 6 in 3643·9 ____________

    b The value of 9 in 285·79 ____________

11. 6 ones + 3 tenths + 8 hundredths = ____________

12. Circle the largest number. Tick the smallest number.

734 902 734 099 734 702

3:4 Extension ☐ out of 7

1. How many legs on 5 cows, 3 horses and 8 chickens? ____________

2. Boxes of 4 globes are packed in this container which has 2 levels. How many globes are there? ____________

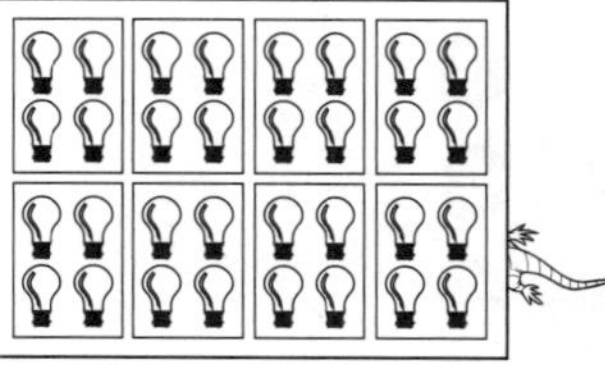

3. If 2 cups of flour make 10 little cakes, how many cups of flour are needed to make 25 little cakes? ____________

4. **a** To half of $1, add 65 cents. ____________

   b From a quarter of $1, take away 20 cents. ____________

5. In our class, there are 26 students. If half of the class have pets, how many students own pets? ____________

6. $4 \times 4 \times 20$ ____________

7. If # means 'plus 7' and ~ means 'subtract 4', then 15 # # # ~ ~ = ____________

Challenge

Create and answer your own questions using the code in question 7 above.

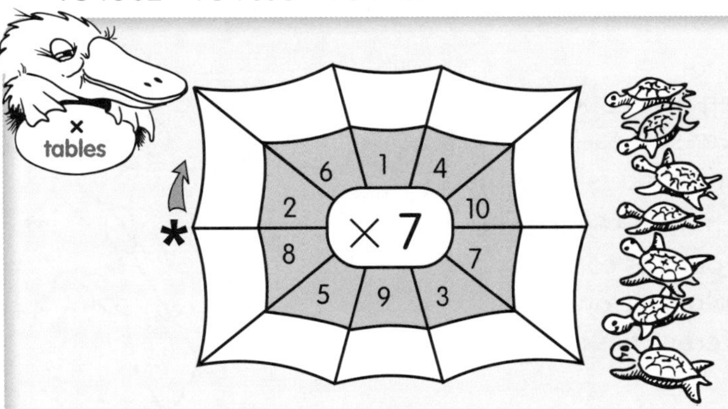

5 × 7	
3 × 7	
7 × 7	
9 × 7	

4:1 out of 20

1. 6 × 2 ______
2. 5 × 3 ______
3. 2 × 2 ______
4. 2 × 4 ______
5. $\begin{array}{r} 563 \\ -\ 353 \\ \hline \end{array}$
6. 4 × ______ = 8
7. 8 × ______ = 16
8. 80 ÷ 10 ______
9. 50 ÷ 10 ______
10. $\begin{array}{r} 782 \\ -\ 356 \\ \hline \end{array}$

11. 3 groups of 4 = ______
12. Complete the first 6 multiples of 3.

    3, ______, ______, ______, ______, ______
13. 800 000 + 45 000 + 300 + 20 =

    ______
14. 8 piles of $3 is worth ______.
15. Write the numeral:

    a 50 000 + 3000 + 200 + 5 ______

    b 20 000 + 300 + 80 + 1 ______
16. What is the perimeter of this rectangle?

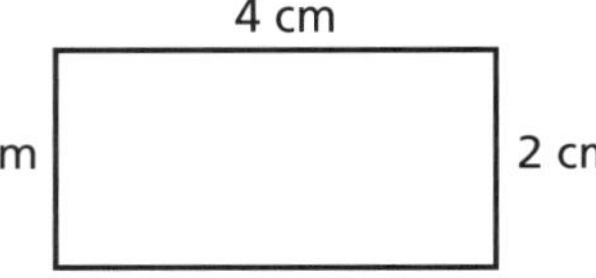

17. The product of 2 and 6 is ______.
18. Write the total value of these coins in decimal form. ______

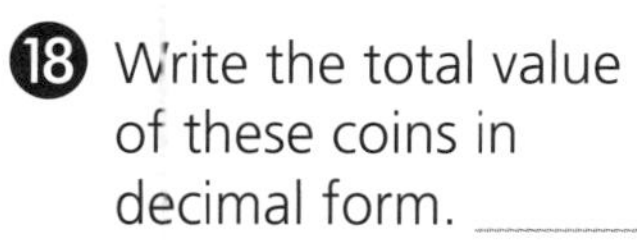

19. Round 8 564 837 to the nearest million. ______
20. Fifty-seven take away twenty-three ______

4:2 out of 18

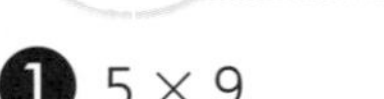

1. 5 × 9 ______
2. 4 × 6 ______
3. 8 × 4 ______
4. 6 × 6 ______
5. $\begin{array}{r} 463 \\ -\ 358 \\ \hline \end{array}$
6. 6 × ______ = 42
7. 5 × ______ = 30
8. Digits in 2 345 321. ______
9. 56 less than 78. ______
10. $\begin{array}{r} 947 \\ -\ 579 \\ \hline \end{array}$

11. The product of 4 and 7 is ______.
12. The value of 15 ten-cent coins. ______
13. Order from largest to smallest.

    35 709 946, 35 738 943, 35 729 499

    ______
14. Write the numeral for:

    a 18 million 426 thousand 364 ______

    b 4 million 200 thousand ______

    c 23 billion ______
15. **a** 354 647 = 350 000 + ______

    b 264 980 = 250 000 + ______
16. 865 000 000 + 450 000 + 34 = ______
17. 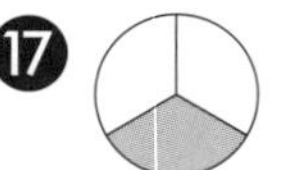 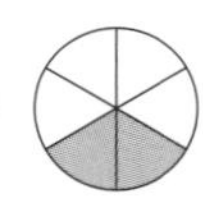 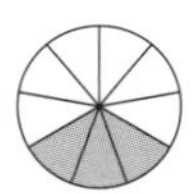

    True (T) or false (F)?

    a $\frac{2}{6} = \frac{1}{3}$ ______ **b** $\frac{2}{6} = \frac{3}{9}$ ______

    c $\frac{2}{3} = \frac{2}{6}$ ______ **d** $\frac{1}{3} = \frac{3}{9}$ ______
18. Half of $2.40 ______

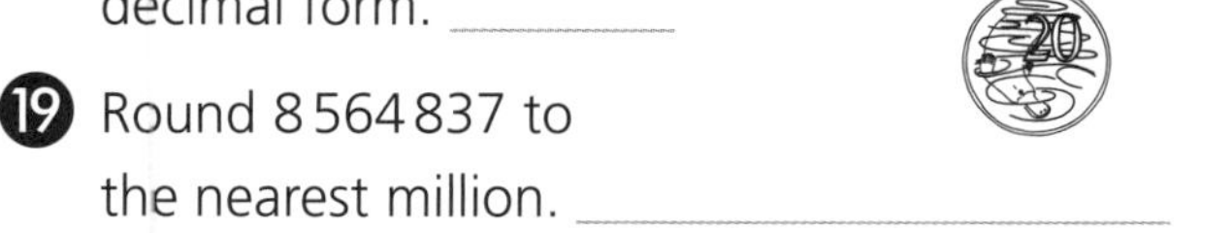

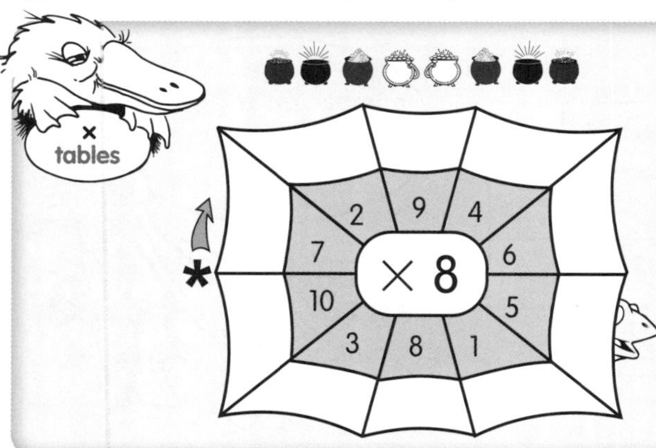

a Greg had 7 sheets of stickers. 8 were on each sheet. How many stickers were there? ______

b Each of the 6 pictures on our ties come in 8 colours. How many different ties do we have? ______

4:3 ☐ out of 9

1. Order from largest to smallest.

   79 026 748, 79 092 736, 89 104 500

   ____

2. Use partitioning and doubling to find:

   a 250 000 + 275 000 ________

   b 150 000 + 189 354 ________

3. Complete the first 6 multiples of 5.

   5, ____, ____, ____, ____, ____

4. 4 children spent $9.

   How much was spent? ________

5. Write the numeral 5 more than:

   a 78 475 038 ________

   b 8 million 74 thousand ________

   c 78 billion and twelve ________

6. 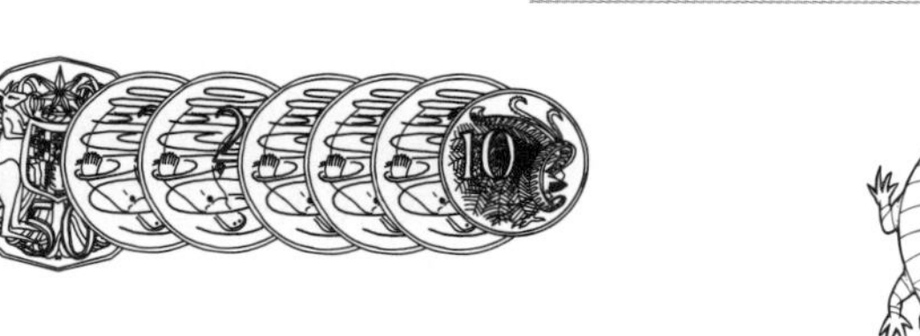

   Write the value of these coins in:

   a cents ________

   b decimal form ________

7. What year was 100 years before 2024? ________

8. Round 3 564 758 to the nearest million. ________

9. Write the mixed number for:

   a $\frac{7}{3}$ ________ **b** $\frac{17}{5}$ ________

4:4 Extension ☐ out of 6

1. 

   a 4 × 32 ________ **b** 5 × 32 ________

2. 14 + 14 + 14 + 14 + 14 ________

3. Use this diagram to find 8 × 34. ________

   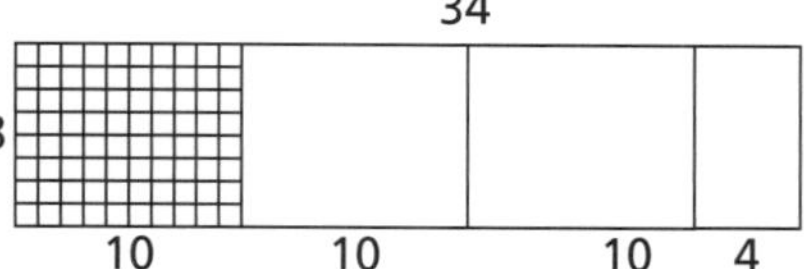

4. Always complete the brackets first.

   a (23 − 17) × 8 ________

   b (7 × 3) + (8 × 4) ________

5. Use partitioning and doubling to find:

   a 800 000 + 875 000 ________

   b 560 000 + 560 354 ________

6. How long would it take Mrs Foster to pay for a washing machine that costs $450 if she pays $50 each week?

   ____

Challenge

List multiples of:

a 2 ____________________

b 3 ____________________

c 4 ____________________

d 5 ____________________

e 6 ____________________

1 × 9		6 × 9	
2 × 9		7 × 9	
3 × 9		8 × 9	
4 × 9		9 × 9	
5 × 9		10 × 9	

5:1 out of 20

1. 3 × 2 ______
2. 5 × 2 ______
3. 2 × ______ = 6
4. 2 × ______ = 14
5. 1 group of 8. ______
6. 3 times 7. ______
7. 4 groups of 2. ______
8. 3 rows of 10. ______
9. Order from largest to smallest.

   405 657   405 836   405 299

   ____
10. Write the numeral six million, twenty-three thousand, four hundred and three. ______
11. **a** 3 × ______ = 18   **b** 18 ÷ 3 = ______
12. 20 000 + 3000 + 500 + 10 + 2 ______
13. Complete the first 6 multiples of 8.

    8, ______, ______, ______, ______, ______
14. $\frac{1}{2}$ and $\frac{\square}{\square}$ makes 1 whole.
15. $\frac{2}{5}$ of the pizza is left.

    What fraction has been eaten? $\frac{\square}{\square}$

16. Colour 2·05 of these hundreds grids.

    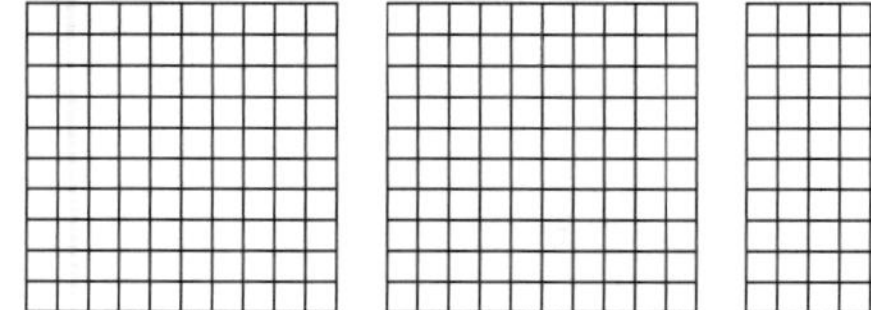
17. Write $6\frac{56}{100}$ as a decimal. ______
18. 0·6, 0·8, 1·0, ______, ______, ______, ______
19. 9 × ▢ = 36   ▢ = ______
20. 11 ÷ 2 = ______ r ______   ○○○○○ ○○○○○○

5:2 out of 21

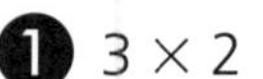

1. 7 × 6 ______
2. 9 × 6 ______
3. 5 × ______ = 40
4. 10 × ______ = 80
5. 3 groups of 9. ______
6. 9 rows of 7. ______
7. 5 times 7. ______
8. 6 × ______ = 60
9. Days in 7 weeks. ______
10. 6 + 3 + ▢ = 12   ▢ = ______
11. 20 ÷ 3 = ______ r ______
12. 6·6, 6·8, 7·0, ______, ______, ______, ______
13. Order from largest to smallest.

    85 700 374   85 078 836   85 809 375

    ____
14. **a** 7 × ______ = 56   **b** 56 ÷ 7 = ______
15. Complete the first 6 multiples of 4.

    4, ______, ______, ______, ______, ______
16. $\frac{1}{4}$ and $\frac{\square}{\square}$ makes 1 whole.
17. $\frac{7}{10}$ of the questions were correct.

    What fraction were not correct? $\frac{\square}{\square}$
18. **a** Colour $\frac{1}{4}$ of this bar.

    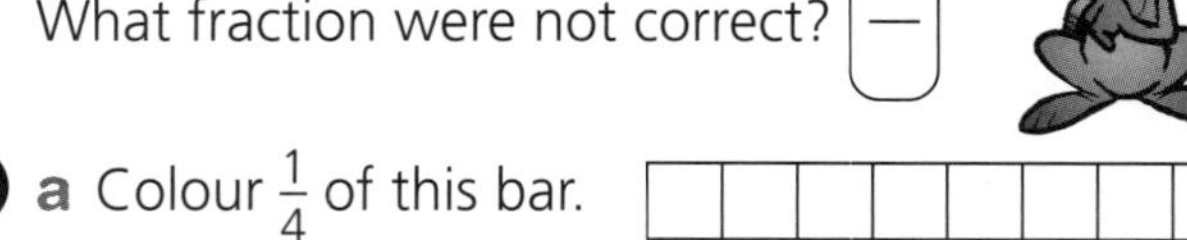

    b What is $\frac{1}{4}$ of 8? ______
19. Write 3·09 as a mixed number. $\frac{\square}{\square}$
20. One dozen is 12. 3 dozen equals ______.
21. **a** Write $\frac{1}{2}$ and $\frac{1}{8}$ on this number line.

    b Circle the larger fraction.

÷ tables

÷ 2	20	12	4	16	14	2	6	18	10	8

÷ 4	20	12	4	16	36	28	40	32	24	8

How many columns of 4?

20 ÷ 4

 • *AUSTRALIAN SIGNPOST MATHS NSW 5 MENTALS* • ISBN 978 0 6557 0912 1

5:3 — out of 12

1. Order these fractions from smallest to largest.

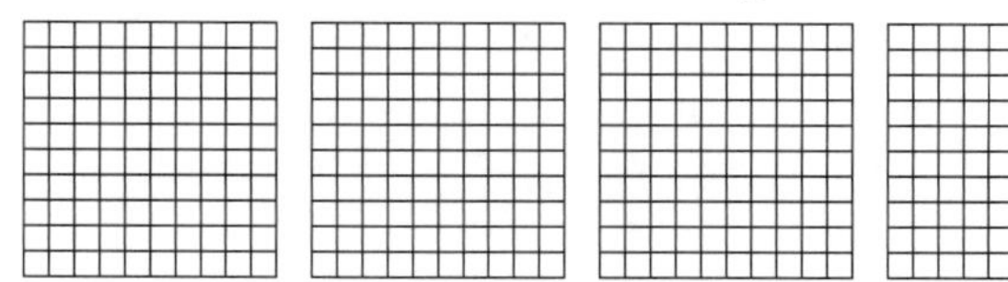

$\frac{1}{4}$, $\frac{1}{10}$, $\frac{1}{5}$ — , — , —

2. Colour 3·75 of these hundreds grids.

3. **a** 9 × ____ = 81 **b** 81 ÷ 9 = ____

4. Order from largest to smallest.
95 384 740, 98 274 834, 92 456 002

5. The value of 6 twenty-cent coins. ____

6. Match each fraction with a part of the hexagon.

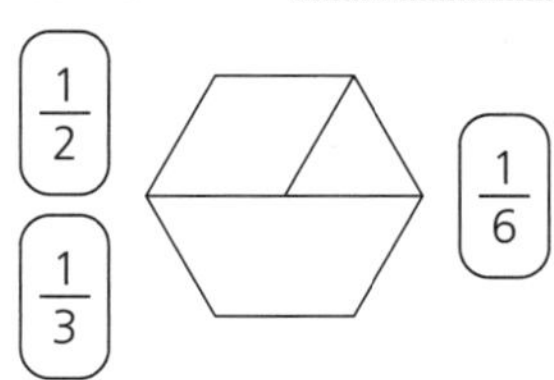

$\frac{1}{2}$ $\frac{1}{3}$ $\frac{1}{6}$

7. Write $3\frac{13}{100}$ as a decimal. ____

8. 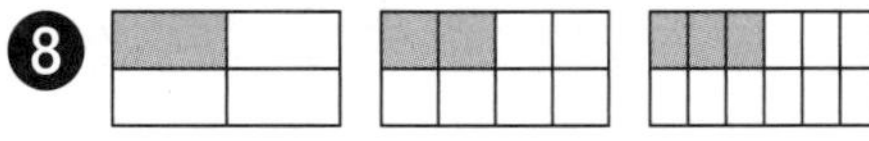

Use the above to write true (T) or false (F).

a $\frac{1}{4} = \frac{2}{8}$ ____ **b** $\frac{1}{4} = \frac{3}{12}$ ____

c $\frac{2}{8} = \frac{3}{12}$ ____ **d** $\frac{3}{4} = \frac{3}{12}$ ____

9. 7 ☐ 7 = 14
The operation for ☐ is: ____

10. The cost of 3 bananas if each costs 95 cents. ____

11. The abbreviation for square centimetres ____

12. 47 ÷ 7 = ____ r ____

5:4 — Extension — out of 6

1. (21 − 11) × (14 − 8) ____

2. If the shaded part has the value given, find the value of the whole.

a 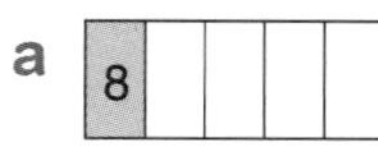

b 

____ ____

3. How many weeks would it take Taya to save $18 for a pet budgie if she saves $1.50 each week? ____

4. Multiples of 7 are 7, 14, 21, … . Which of 94, 96 and 98 is the sum of two multiples of 7? ____

5. Seven cubes of side length 1 cm are glued as shown. How many faces has the shape? ____

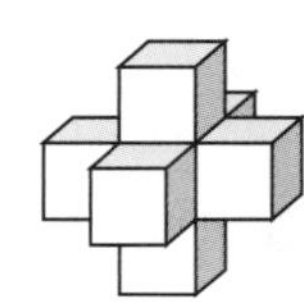

6. $\frac{4}{10}$ of 10 is ____. $\frac{4}{10}$ of 40 is ____.

Challenge

Represent and label two mixed numbers.

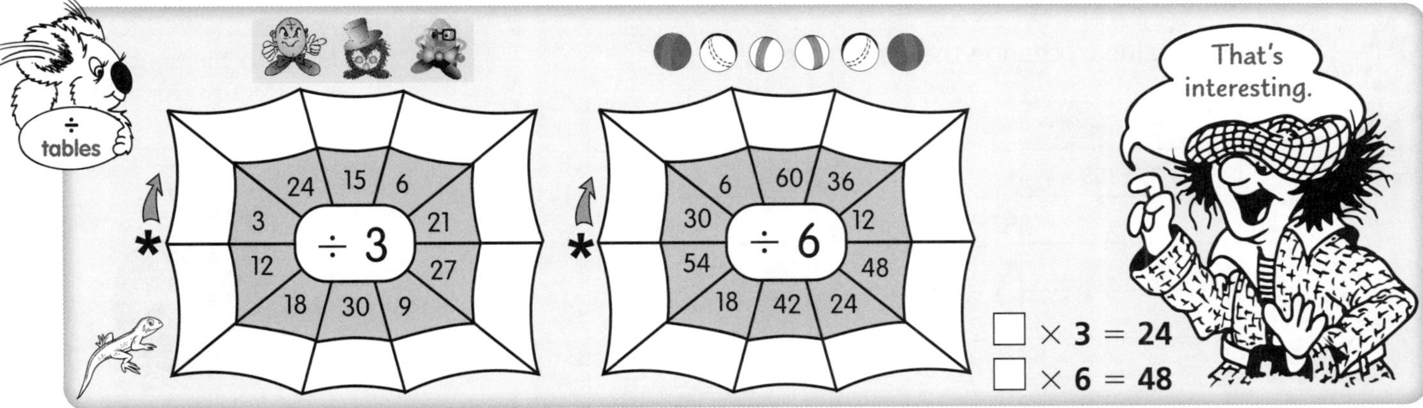

6:1 out of 19

1. 2 × ______ = 10
2. 10 × ______ = 30
3. 60 ÷ 6 ______
4. 90 ÷ 10 ______
5. 34 more than 25. ______
6. 63 take away 21. ______
7. Digits in 3 546 432. ______
8. Double 10 × 4. ______
9. Write the decimal shown by the grey blocks. ______

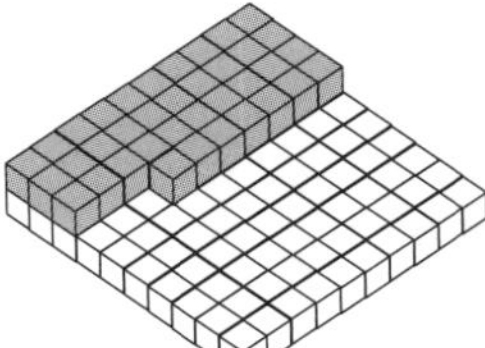

10. $\frac{3}{10}$ and $\frac{\square}{\square}$ makes 1 whole.
11. a 2 × ______ = 8 b 8 ÷ 2 = ______
12. 28 + 6 = 30 + ______ = ______
13. Write the mixed number and decimals shaded.

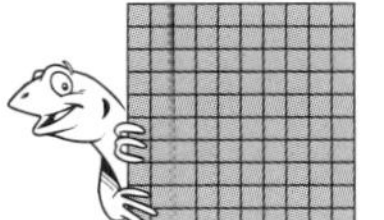
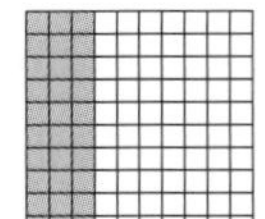
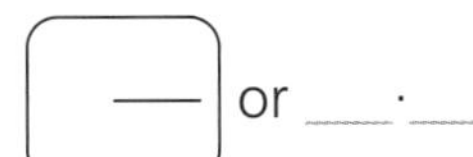

$\frac{\square}{\square}$ or ___·___

14. How many kilometres in 3000 m? ______
15. Write 5·37 as a mixed number. $\frac{\square}{\square}$
16. Match each fraction to the correct decimal.

$\frac{6}{100}$	0·1
$\frac{1}{10}$	0·7
$\frac{7}{10}$	0·06

$\frac{13}{100}$	0·13
$\frac{8}{10}$	0·74
$\frac{74}{100}$	0·8

17. Complete these equivalents:

0·25	$\frac{\quad}{100}$	%

0·50	$\frac{\quad}{100}$	%

18. a 16 ÷ 8 = ______ b 20 ÷ 2 = ______
19. 27 ÷ 5 = ______ r ______

6:2 out of 20

1. 6 × ______ = 54
2. 9 × ______ = 63
3. 72 ÷ 9 ______
4. 81 ÷ 9 ______
5. 4 squared. ______
6. 70 divided by 7. ______
7. 8 times 1000. ______
8. 5 groups of 10. ______
9. List the least number of notes and coins needed to make $231.25. ______

______

10. $\frac{1}{3}$ of the container was full. What fraction was empty? $\frac{\square}{\square}$

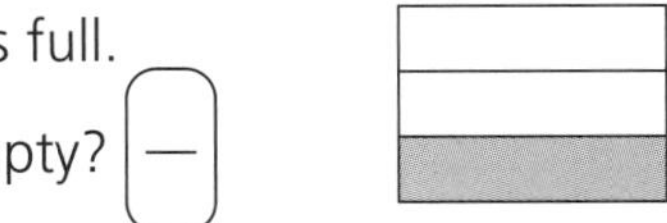

11. 8 squared means 8 × 8. 8 squared is ______.
12. Write 9·94 as a mixed number. $\frac{\square}{\square}$
13. Use partitioning and doubling to find:
    a 350 000 + 369 000 ______
    b 420 000 + 437 354 ______
14. a Colour $\frac{1}{4}$ of this length.

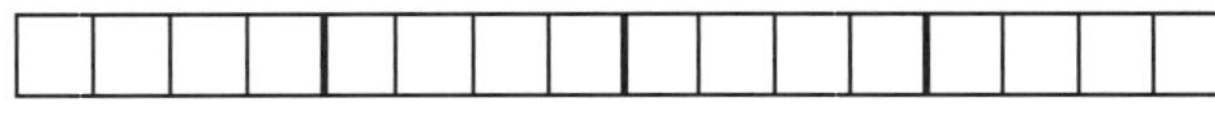

    b What is $\frac{1}{4}$ of 16? ______
    c What fraction of the bar is not coloured? $\frac{\square}{\square}$
15. a Write $\frac{1}{2}$ and $\frac{1}{4}$ on this number line. 0 ——— 1
    b Circle the larger fraction.
16. How many metres in 6 km? ______
17. Write 8 tenths as a decimal. ______
18. 6·3, 6·4, 6·5, ______, ______, ______, ______
19. Write the decimal for $4\frac{29}{100}$. ______
20. 39 ÷ 6 = ______ r ______

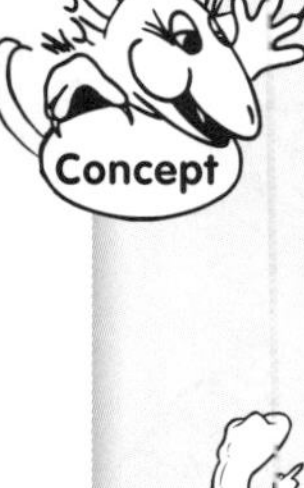

Use place value to change these numbers.

a

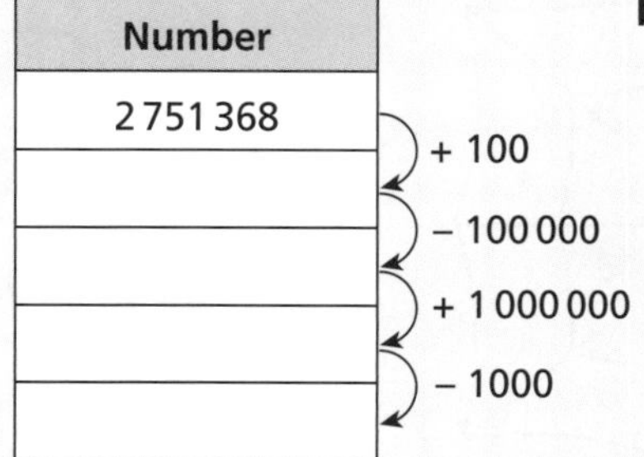

Number	
2 751 368	+ 100
	– 100 000
	+ 1 000 000
	– 1000

b

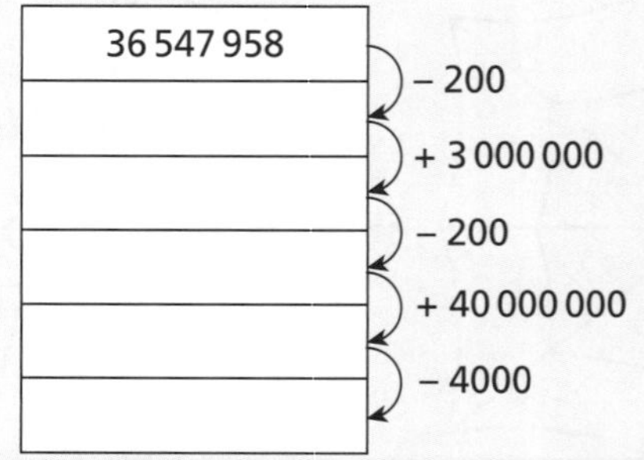

36 547 958	– 200
	+ 3 000 000
	– 200
	+ 40 000 000
	– 4000

 ISBN 978 0 6557 0912 1

6:3 ☐ out of 12

1. 50 ÷ 8 = ______ r ______
2. $\frac{4}{5}$ and $\frac{\square}{\square}$ makes 1 whole. ☐☐☐☐☐
3. $\frac{1}{4} + \frac{1}{4} + \frac{1}{4} = \frac{\square}{\square}$ ☐☐☐☐
4. How many kilometres in 6000 m? ______
5. Would you use m, km, mm or cm to measure the length of a:
   a house? ______ b finger? ______
   c road? ______ d ring? ______
6. a Shade 20 out of 100.
   b What fraction would not be shaded? $\frac{\square}{\square}$
   c What percentage is shaded? ______

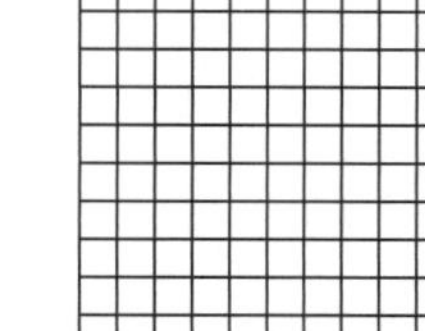

7. a 8 × ______ = 72 b 72 ÷ 8 = ______
8. Write the mixed number and decimals shaded.

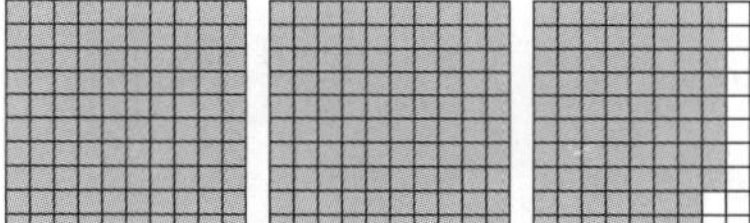

or ___.___

9. A minibus carries 8 people. How many people can fit on 4 minibuses? ______
10. Aziz had 52 marbles. He lost 12 on Monday, 10 on Wednesday and 7 on Thursday. How many did he have left? ______

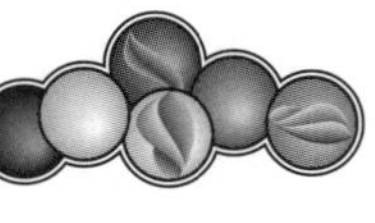

11. What is the value of the 4 in 3 653 493? ______
12. Complete the label.

______ past ______

6:4 ☐ out of 7 Extension

1. Use partitioning and doubling to find:
   a 550 000 + 586 000 ______
   b 550 000 + 551 464 ______
2. There are 12 chocolates on each of the two layers in a box. How many chocolates are in 3 boxes? ______
3. What is the value of 70 gold bars worth $1000 each? ______
4. a What two numbers add to give 13 and multiply to give 30? ______
   b What two numbers add to give 14 and multiply to give 24? ______
5. If # means 'times 2' and ^ means 'divide by 3', then 12 # # ^ = ______
6. If you look at this watch once every minute from 8:40 pm until 9:00 pm, how many fives would you see?

______

7. a 6 squared plus 9 squared ______
   b Add 28 to 5 squared. ______

Challenge

Write as much as you can about the number 6 534 108.

______

______

______

______

______

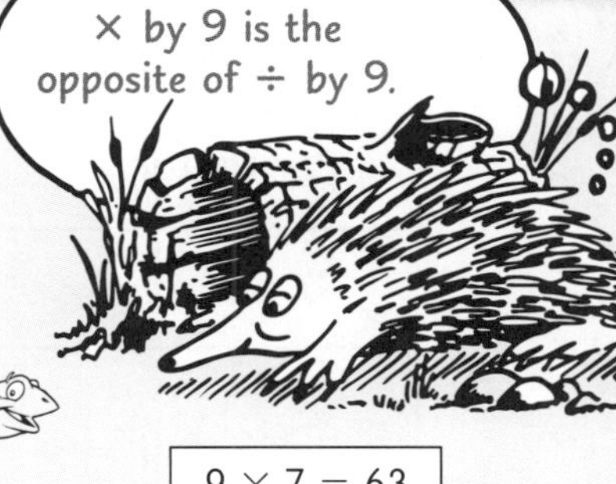

9 × 7 = 63
63 ÷ 9 = 7

7:1

out of 19

1. $14 \div 2$ ______
2. $21 \div 3$ ______
3. $50 \div 10$ ______
4. $18 \div 3$ ______
5. Add 23 to 64. ______
6. 6 multiplied by 5. ______
7. 6 groups of 5. ______
8. Halve 64. ______
9. a $6 \div 10$ 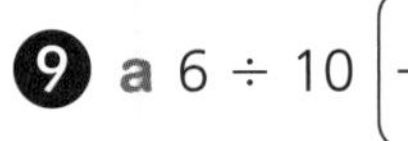

   b $2 \div 5$ 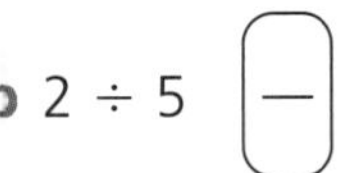

10. How many metres in 8 km? ______
11. Change $\frac{9}{4}$ to a mixed numeral. 
12. Complete these equivalents.

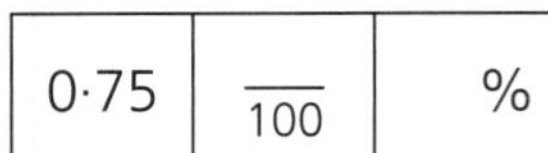

0·75	$\frac{}{100}$	%

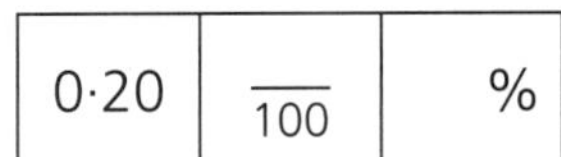

0·20	$\frac{}{100}$	%

13. Change $1\frac{3}{4}$ to an improper fraction. 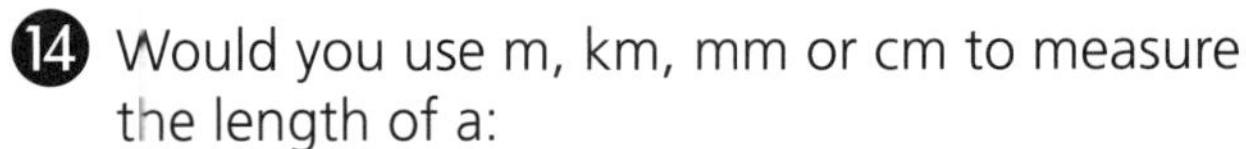
14. Would you use m, km, mm or cm to measure the length of a:

    a driveway? ______ b pea? ______

    c pen? ______ d river? ______
15. What is the perimeter of this rectangle? ______

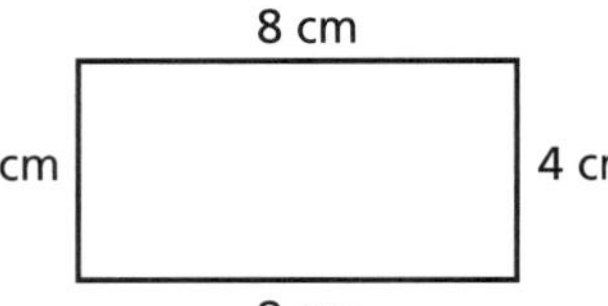

16. Write zero point two as a decimal. ______
17. How many 20c coins make $2? ______
18. 0·43 = ______ out of ______ = ______%
19. Write the numeral for 5 billion. ______

7:2

out of 17

1. $10 \div 10$ ______
2. $63 \div 9$ ______
3. $56 \div 8$ ______
4. $54 \div 6$ ______
5. 42 shared by 7. ______
6. 64 divided by 8. ______
7. $\frac{1}{4}$ of 20. ______
8. $\frac{1}{4}$ of 16. ______
9. Convert $\frac{11}{3}$ to a mixed number.
10. Use the compensation strategy to find:

    a 145 + 198 ______ b 325 + 297 ______
11. Write each as kilometres and metres.

    a 4638 m = ______ km ______ m

    b 8305 m = ______ km ______ m
12. Convert $3\frac{4}{10}$ to an improper fraction.
13. Write the part shaded as:

    a ______ hundredths

    b a fraction

    c a percentage ______

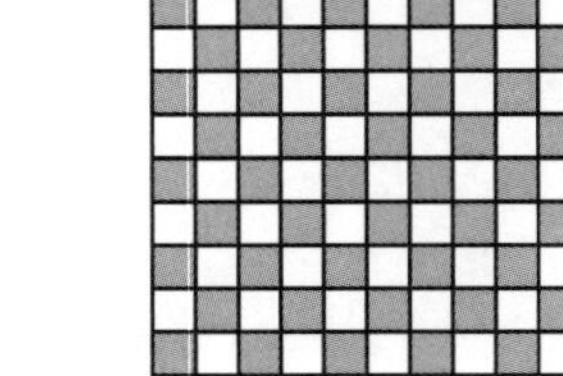

14. I travelled 30 km/h. How far did I travel in 4 hours? ______
15. Complete the following.

a

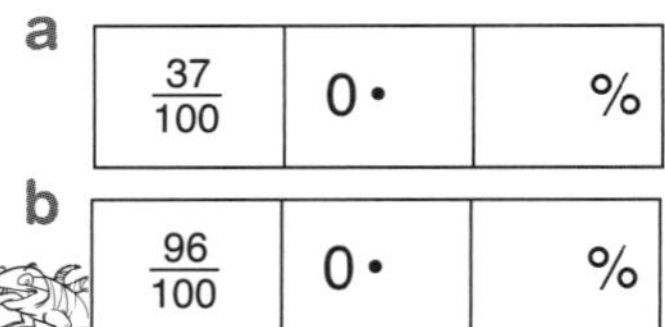

$\frac{37}{100}$	0•	%

b

$\frac{96}{100}$	0•	%

16. Halves in $7\frac{1}{2}$. ______

17. What is 3 more than 83 456 649? ______

Measure

Find the perimeter of each shape.

a Perimeter = ______ cm

b Perimeter = ______ cm

7:3 ☐ out of 10

1. Convert $\frac{15}{4}$ to a mixed number. ☐ $\frac{\square}{\square}$

2. a $7 \div 8$ ☐ b $2 \div 10$ ☐

3. Write each length in metres.
   a 8 km 156 m = ________ m
   b 7 km 365 m = ________ m

See page 85 for help.

4. I travelled 50 km/h.
   How far did I travel in 6 hours? ________

5. Complete:
   a ________ hundredths
   b ________ out of ________
   c 0 · ________
   d ________ %

6. Convert $7\frac{3}{10}$ to an improper fraction. ☐

7. The perimeter of this shape.
   ____

8. a Colour 46% of this square.
   b What percentage is not coloured? ________

9. a Round 7624 correct to the nearest thousand. ________
   b Round 9654 correct to the nearest thousand. ________

10. a Metres in 6 kilometres? ________
    b Grams in 8 kilograms? ________

7:4 Extension ☐ out of 5

1. The cost of:

HAMBURGERS
\$2.70 each, or
2 for \$5.00

   a 4 hamburgers ________
   b 12 hamburgers ________

2. a $(4 \times 9) + (5 \times 5)$ ________
   b $(24 - 18) \times 5$ ________

3. a $\frac{1}{2}$ of 20 ________
   b $\frac{1}{4}$ of 20 ________
   c $\frac{1}{5}$ of 20 ________
   d $\frac{1}{10}$ of 20 ________

4.

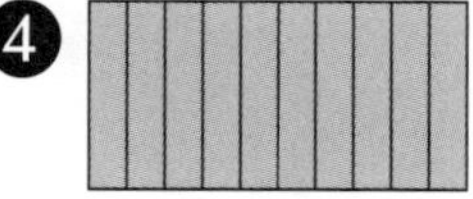

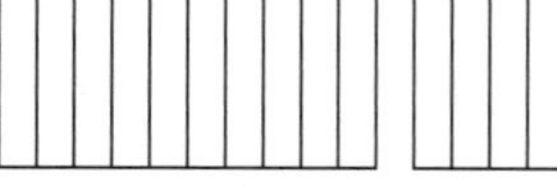

   a $1 \div 2$ ________ b $3 \div 2$ ________

5. Complete this pattern and write the rule.
   206, 308, 410, ________, ________ Rule: ________

Challenge

Complete this table.

Fraction	Decimal	Percentage
$\frac{20}{100}$	0·20	20%
	0·41	
		86%
$\frac{38}{100}$		

Turn to ID card C on page 8.
Give the answers for these numbers.

(19) 3 out of ______ (20) $\frac{\square}{10}$ shaded

(21) ______ tenths shaded (22) 0·______ shaded

(23) $\frac{\square}{100}$ shaded (24) 0·______ shaded

(26) 19 out of ______ (27) ______ % shaded

(28) ______ · ______ (30) $0{\cdot}3 = \frac{\square}{10}$, $0{\cdot}30 = \frac{\square}{10}$

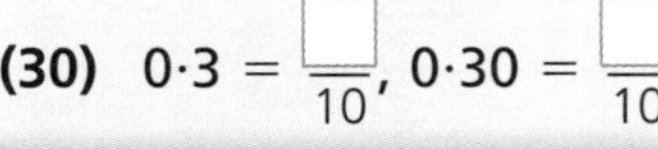

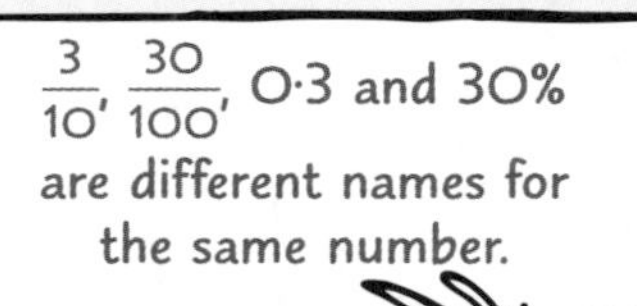

8:1

out of 15

1. 56 + 50 ______
2. 52 + 56 ______
3. 12 ÷ 4 ______
4. 28 ÷ 4 ______
5. 7 groups of 5. ______
6. 8 rows of 3. ______
7. Add 45 and 35. ______
8. Multiply 4 and 6. ______
9. Convert $\frac{19}{5}$ to a mixed number.

10. $\frac{1}{4} + \frac{1}{4}$ ☐☐☐☐ $\frac{\square}{\square}$
11. What fraction is shaded? ______

12. Shade $\frac{8}{12}$ of this shape.

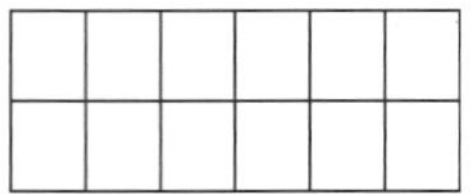

13. Share 36 cakes among 4 children. ______ each
14. **Books read in a week**

	Tom	Pam	Cole	Deb	Tim
Pictures	📖📖	📖📖📖	📖📖📖📖📖	📖📖📖📖	📖

(Scale: 2, 4, 6, 8, 10)

📖 = 2 books

a How many books were read by Tom and Deb altogether? ______

b How many books read, does one picture stand for? ______

c How many books were read altogether? ______

15. a Tens in 5 364 545? ______

b Hundreds in 3 045 345? ______

8:2

out of 17

1. 156 + 199 ______
2. 465 + 297 ______
3. 83 − 67 ______
4. 51 − 38 ______
5. 7 squared. ______
6. 48 divided by 6. ______
7. 32 shared by 4. ______
8. 6 × ______ = 30
9. Write each improper fraction as a mixed number.

a $\frac{7}{4}$ ______ b $\frac{5}{3}$ ______

10. Write as an improper fraction.

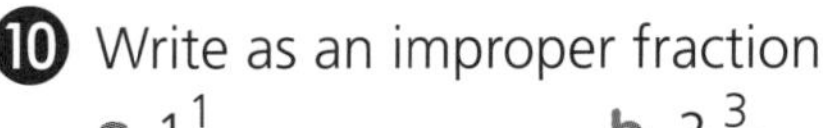

a $1\frac{1}{4}$ ______ b $2\frac{3}{10}$ ______

11. a What is the perimeter of this shape? ______

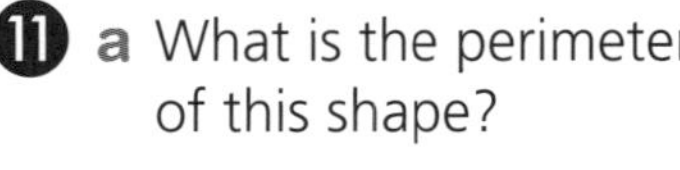

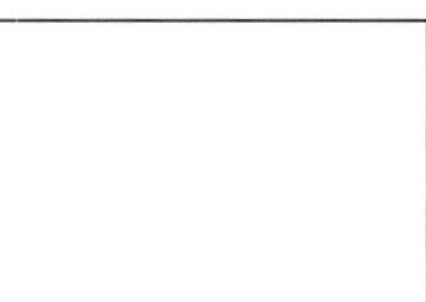

b What is the area of this shape? ______

12. a Shade $\frac{3}{6}$ of this shape.

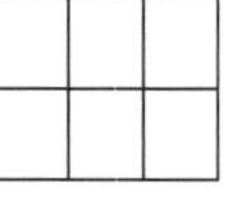

b $\frac{3}{6} - \frac{2}{6}$ ______

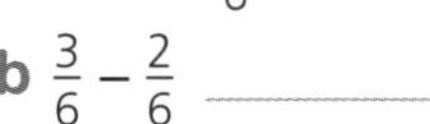

13. Colour $2\frac{1}{4}$ rectangles.

☐ ☐ ☐ ☐

14. Write 65c in decimal form. ______
15. Write the numeral:

a fifty-eight million, five hundred and thirty-one thousand, four hundred and three ______

b seventy-nine point two ______

c thirty-five point six five ______

16. How many sides has a quadrilateral? ______
17. 15, 24, 33, ______, ______, ______, ______

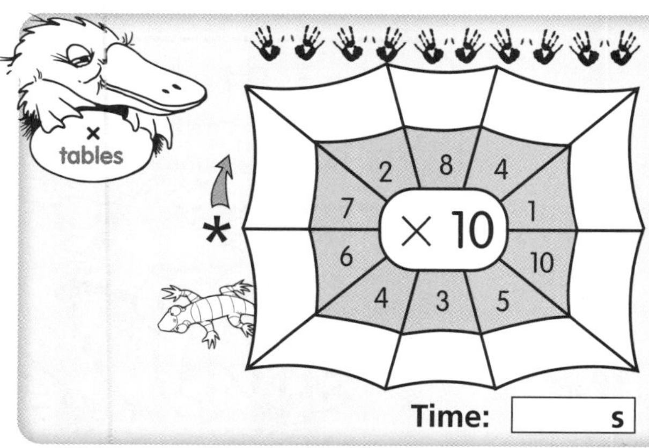

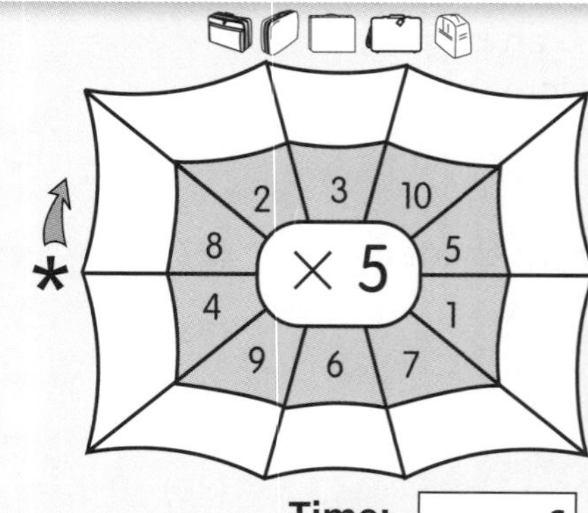

8:3 ☐ out of 7

1 **a** What is the perimeter of this shape? ____

b What is the area of this shape? ____

2 Write 35 million six hundred. ____

3 Write 9 tenths as a decimal. ____

4 My ruler is 31 cm long and 4 cm wide. What is the:

a perimeter of my ruler? ____

b area of my ruler? ____

5

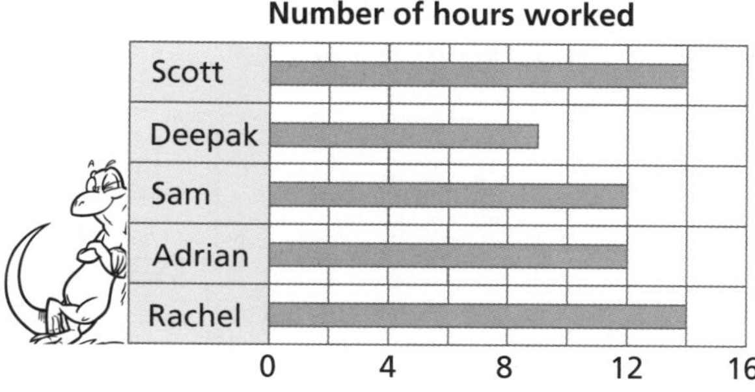

a For how long did Adrian work? ____

b How many hours were worked altogether? ____

6 Write the number:

a after 56 734 629 ____

b before 39 475 280 ____

c before 19 273 400 ____

7 **a** Colour 3 sevenths of this rectangle.

b Colour $\frac{2}{7}$ of the rectangle.

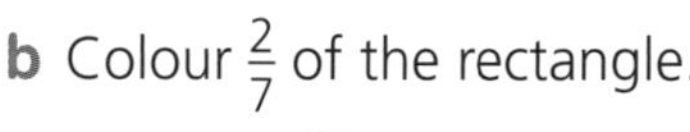

c $\frac{3}{7} + \frac{2}{7} = \frac{\square}{\square}$ **d** $\frac{5}{7} - \frac{2}{7} = \frac{\square}{\square}$

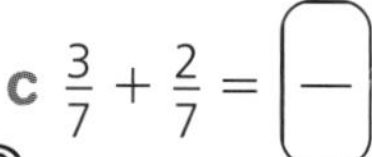

8:4 Extension ☐ out of 5

1 If the shaded part has the value given, find the value of the whole.

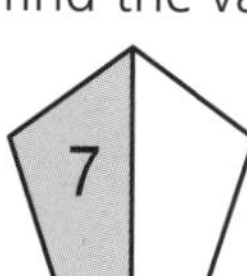

____

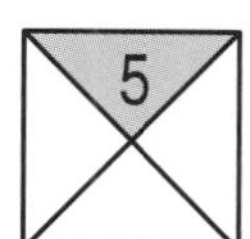

____

2 I spent half of my money on Thursday and one quarter of it on Friday. I had $2.50 left. How much did I start with?

____

3 **a** $10\frac{1}{2} - 7$ ____

b $8\frac{1}{2} - 2$ ____

4 How much carpet would I need for my bedroom floor, which is 3 m by 6 m and my lounge room floor, which is 4 m by 8 m? ____

5 Which is larger, 9×4 or 8×6? ____

Challenge

Write number sentences for addition of fractions, with a denominator of 8, e.g. $\frac{1}{8} + \frac{4}{8} = \frac{5}{8}$.

ID Card A

Turn to ID Card A on page 6.

Give the answer for these numbers.

(1) ____ **(2)** ____ **(3)** ____

(4) ____ **(5)** ____ **(6)** ____

(7) ____ **(8)** ____ **(10)** ____

(13) ____

 • *AUSTRALIAN SIGNPOST MATHS NSW 5 MENTALS* • ISBN 978 0 6557 0912 1

9:1 ☐ out of 14

1. 9 + 6 ____
2. 7 × 3 ____
3. 18 − 17 ____
4. 5 × 4 ____
5. $\begin{array}{r} 676 \\ -\,452 \\ \hline \end{array}$
6. 8 times 2. ____
7. 21 plus 7. ____
8. Add 36 and 4. ____
9. 30 divided by 5. ____
10. $\begin{array}{r} 625 \\ -\,481 \\ \hline \end{array}$

11. a What fraction is shaded? ☐

b What fraction is not shaded? ☐

c $\frac{2}{6} + \frac{4}{6} =$ ☐

12. A 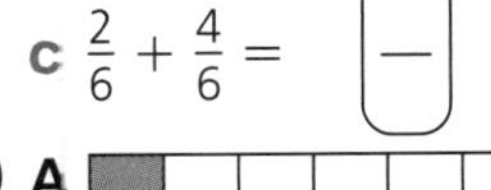 B 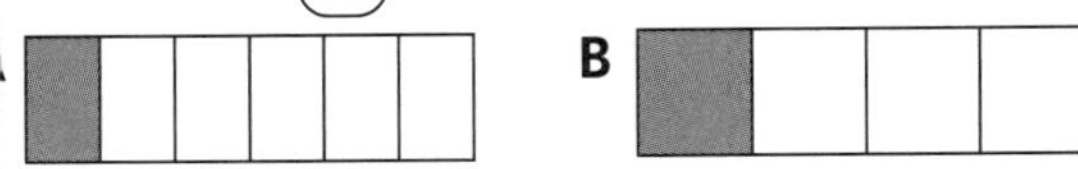

Which of these fractions would:

a have the larger denominator? ____

b be the smaller fraction? ____

13. a What fraction of this shape is shaded? ____

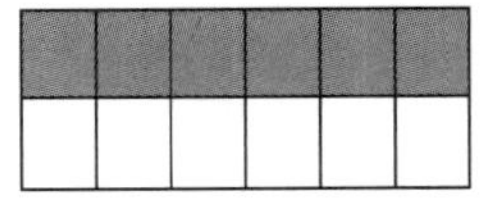

b $\frac{6}{12} - \frac{3}{12}$ ____

14. On this place-value chart, write:

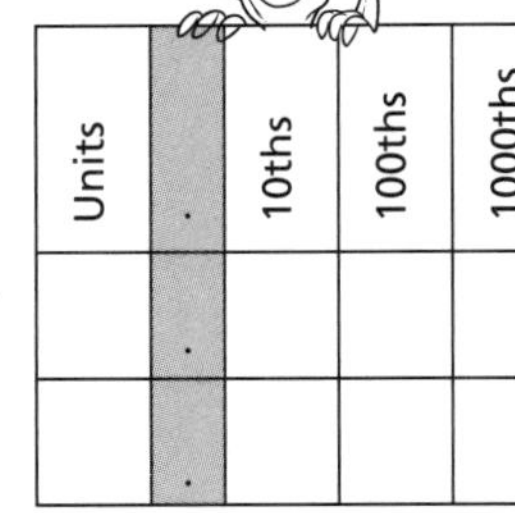

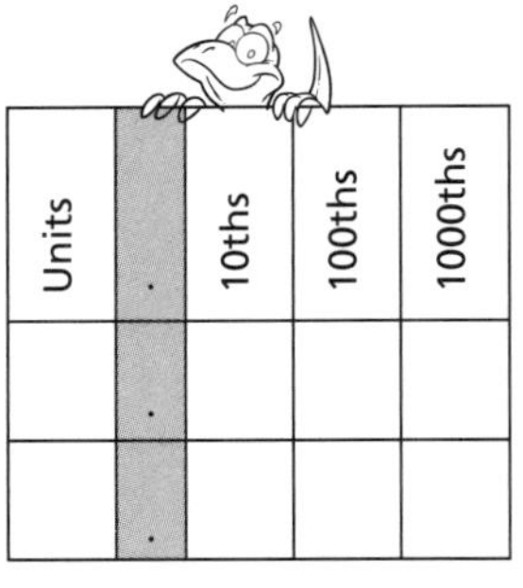

Units	.	10ths	100ths	1000ths
	.			
	.			

a six point three one four

b eight point nine two seven.

9:2 ☐ out of 16

1. 56 − 16 ____
2. 7 × 5 ____
3. 89 − 35 ____
4. 152 + 256 ____
5. $\begin{array}{r} 364 \\ +\,462 \\ \hline \end{array}$
6. 8 multiplied by 7. ____
7. 53 take away 49. ____
8. 16 divided by 4. ____
9. Groups of 6 in 36. ____
10. $\begin{array}{r} 863 \\ -\,384 \\ \hline \end{array}$

11. $\frac{1}{4} + \frac{1}{4} + \frac{1}{4} + \frac{1}{4}$ 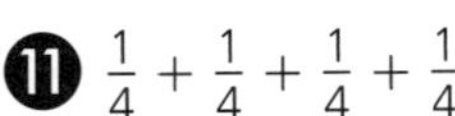 ____

12. My wall is 5 metres long and 3 metres high. What is the area of my wall?

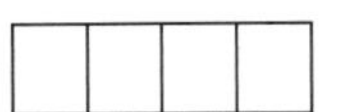

13. a

Area = ____

b 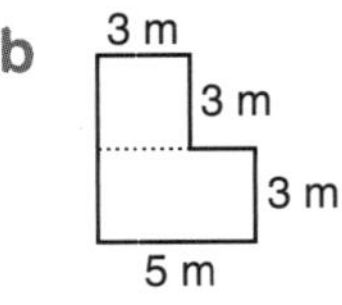

Area = ____

14. Jessica has 4 dolls, 12 games and 8 spinning tops. Complete the column graph to show the number of each toy.

Jessica's toys

Dolls							
Games							
Tops							
	0		4		8		12

15. Write a number that has:

a 5 thousandths ____

b 4 hundredths ____

c 9 tenths ____

16. Write the decimal for 34 and 56 hundredths. ____

Concept: Using a graph

Crosses or dots can be used on a graph.

Draw a graph of these colour choices made by students.

B, R, R, G, P, B, O, R, B, R, G, P, O, R

G, B, P, O, Y, Y, R, B, O, G, G, B, B, P

a Which is the favourite colour? ____

b How many students chose green? ____

c How many more chose red than pink? ____

B = blue
G = green
O = orange
P = pink
R = red
Y = yellow

 • *AUSTRALIAN SIGNPOST MATHS NSW 5 MENTALS* • ISBN 978 0 6557 0912 1

9:3 ☐ out of 6

1. $\frac{1}{4} + \frac{3}{4}$  ______

2. My post-it note is a square 8 cm long. What is the:
   a perimeter? ______
   b area? ______

3. 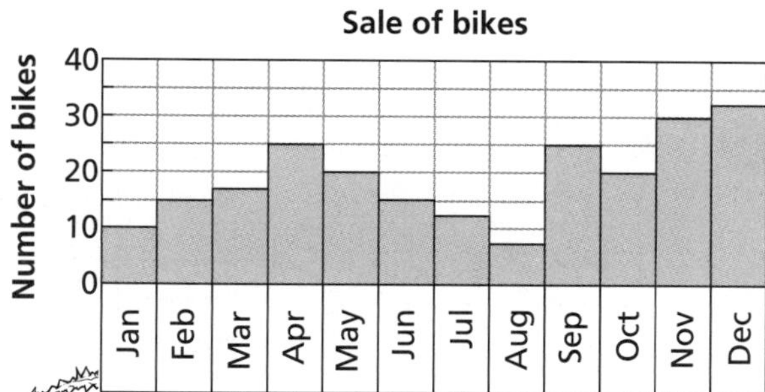

   a Which month had the most sales? ______
   b Why would this month have the most sales?
   ______
   c How many sales in spring? ______

4. On this place-value chart, write:
   a nine point six four two
   b three point eight one
   c six point zero seven nine

Units	.	10ths	100ths	1000ths
	.			
	.			
	.			

5. Write the decimal for 67 and 471 thousandths. ______

6. Would it be better to use cm² or m² to measure the area of a handball court? ______

Extension 9:4 ☐ out of 7

1. What fraction of this shape is shaded? ______
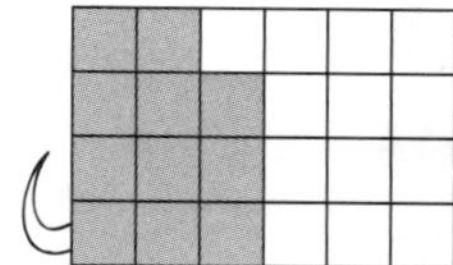

2. My wall is 10 m wide and 5 m high. There is a window on this wall that is 3 m wide and 2 m high. How much of the wall is not covered by the window? ______

3. Months in $7\frac{1}{4}$ years? ______

4. 'am' means: ______

5. Circle the larger decimal.
   56 and 475 thousands **or** 89 and 99 hundredths

6. I was at the shops from 4:26 pm to 5:54 pm. For how long was I at the shops? ______

7. a I want to tile an area around my pool which is 12 m by 9 m. The pool is 4 m by 9 m.
   What area will be tiled? ______
   b If I pay $12 per square metre for tiles, how much will it cost me? ______

Challenge

Draw and label fractions that add to make a whole.

Area and perimeter (challenge)

The area of this shape is 100m².

What is the perimeter?

______

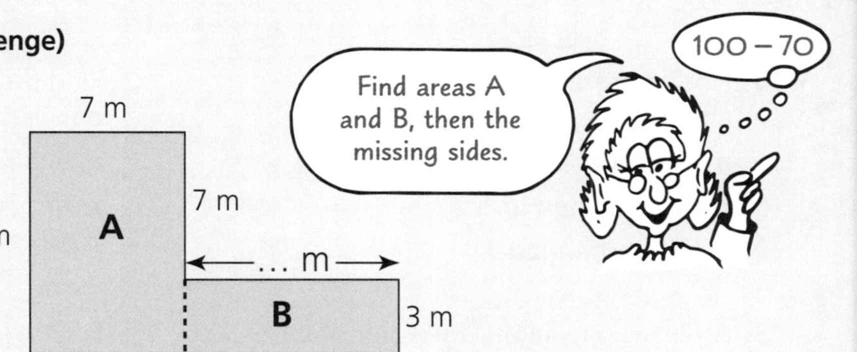

10:1 ☐ out of 16

1. 65 − 16 ______
2. 47 + 98 ______
3. 420 − 143 ______
4. 162 − 99 ______
5. $\begin{array}{r} 792 \\ +\,148 \\ \hline \end{array}$
6. Multiply 8 and 3. ______
7. 205 plus 207. ______
8. Groups of 7 in 70. ______
9. 24 divided by 4. ______
10. $\begin{array}{r} 853 \\ -\,278 \\ \hline \end{array}$

11. | | | | |
|---|---|---|---|

a $\frac{3}{4} - \frac{2}{4}$ ______ b $\frac{2}{4} - \frac{1}{4}$ ______

12. What is the area of this battleaxe block?

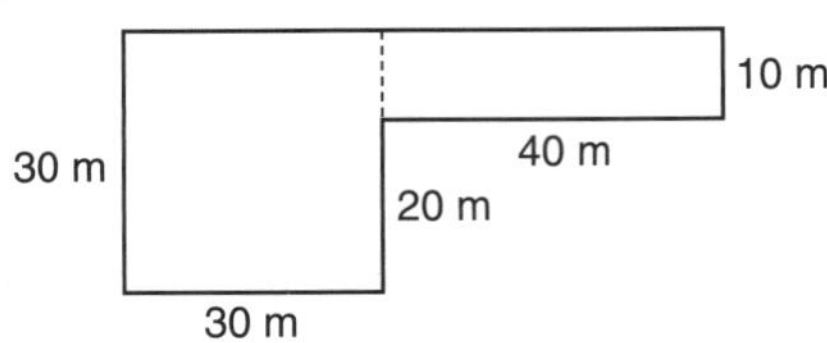

______

13. Make a tally for the animals in this farmyard.

Chickens	
Pigs	
Cows	

14. a Fraction shaded = $\frac{\square}{\square}$

b Fraction not shaded = $\frac{\square}{\square}$

c $\frac{3}{4} + \frac{\square}{\square}$ make 1 whole.

15. Write the decimal for 46 hundredths. ______
16. Round 574 to the nearest hundred. ______

10:2 ☐ out of 15

1. 5 × 8 ______
2. 246 + 337 ______
3. 183 − 35 ______
4. 9 × 6 ______
5. $\begin{array}{r} 745 \\ +\,257 \\ \hline \end{array}$
6. Double 17. ______
7. 25 divided by 5. ______
8. 54 minus 18. ______
9. Halve 52. ______
10. $\begin{array}{r} 975 \\ -\,568 \\ \hline \end{array}$

11. Use the table to complete this column graph.

Team	Number
Kings	12
Eels	8
Rams	20

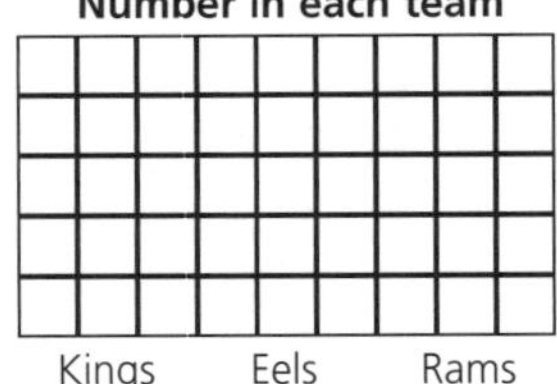

12. 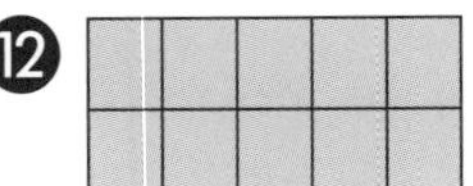

a $\frac{2}{10} + \frac{5}{10} =$ ______ b $\frac{4}{10} + \frac{4}{10} =$ ______

c $\frac{3}{10} + \frac{7}{10} =$ ______ d $\frac{2}{10} + \frac{6}{10} =$ ______

13. On this place-value chart, write:

a four point three four

b seven point nine two

Units	.	10ths	100ths	1000ths
	.			
	.			

14. On this number line, place a dot at 2·73.

15. Use constant difference to find 435 − 227. ______

251, 351, 451

1 Find the total of the numbers above if they are:

a first rounded to the nearest 100 ______

b first rounded to the nearest 10 ______

c left unchanged ______

249, 349, 449

2 Find the total of the numbers above if they are:

a first rounded to the nearest 100 ______

b first rounded to the nearest 10 ______

c left unchanged ______

3 **a** (251+351+451)−(249+349+449) ______

b (300+400+500)−(200+300+400) ______

 • *AUSTRALIAN SIGNPOST MATHS NSW 5 MENTALS* • ISBN 978 0 6557 0912 1

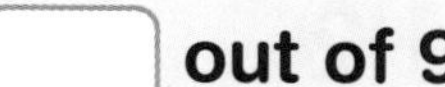

10:3 — out of 9

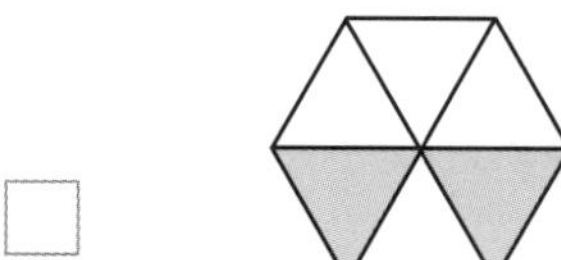

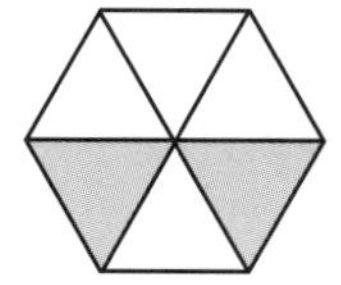

1. a Fraction shaded = $\frac{\square}{\square}$
   b Fraction not shaded = $\frac{\square}{\square}$
   c $\frac{2}{6} + \frac{\square}{6}$ make 1 whole.
2. How many pictures are needed to represent:

   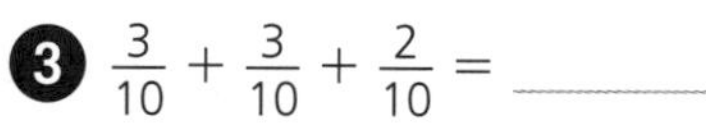

   a 90 ships? ______ b 120 ships? ______
3. $\frac{3}{10} + \frac{3}{10} + \frac{2}{10} =$ ______
4. Write the decimal for:
   a 36 and 56 hundredths ______
   b 83 and 264 thousandths ______
5. On this number line, place a dot at 7·58.

6. a 8 weeks = ______ days
   b 180 minutes = ______ hours
   c 60 years = ______ decades
7. How many hours from the time on this watch until 3 pm? ______

8. The time:
   a 6 hours after noon ______
   b 3 hours before midnight ______
9. Use levelling (take from one number and add it to the other number) to find:
   a 69 + 34 ______ b 497 + 365 ______

10:4 — out of 7

Extension

1. Which is larger, 12 × 8 or 14 × 7? ______
2. a How many minutes in ten hours? ______
   b How many minutes in one day? ______
   c 2010 years = ______ decades
3. Look for compatible numbers to find:
   a 348 + 556 ______ b 497 + 893 ______
4. Use place value to find:
   a 546 − 328 ______ b 783 − 464 ______
5. If this pattern continued, what would the 35th shape look like?
   ▲, ●, ■, △, ○, □, ▲, ●, ______
6. Use levelling (take from one number and add it to the other number) to find:
   a 3546 + 993 ______
   b 2859 + 887 ______
7. Use constant difference (add to or subtract from both numbers) to find:
   a 9183 − 4945 ______
   b 4785 − 2428 ______

Challenge

Write questions that are equal to:

a 567 + 354	b 265 + 465	c 1045 + 3548
= ______	= ______	= ______
= ______	= ______	= ______
= ______	= ______	= ______
= ______	= ______	= ______
= ______	= ______	= ______

The jump strategy

We can do this on a number line or as a mental strategy.

Example:

347 + 34

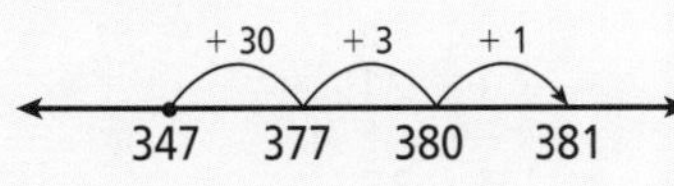

Use the jump strategy for these:

a 778 − 426 = ______

778

b 649 + 32 = ______

649

11:1 ☐ out of 17

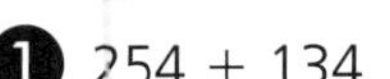

1. 254 + 134 ______
2. 623 + 134 ______
3. 465 − 153 ______
4. 536 − 325 ______
5.
   456
   + 367
6. 354 + 8 ______
7. 638 + 9 ______
8. 376 + 99 ______
9. 365 + 97 ______
10.
   736
   + 198

11.  Round 6439 to the nearest:
    a ten ______ b thousand ______
12. 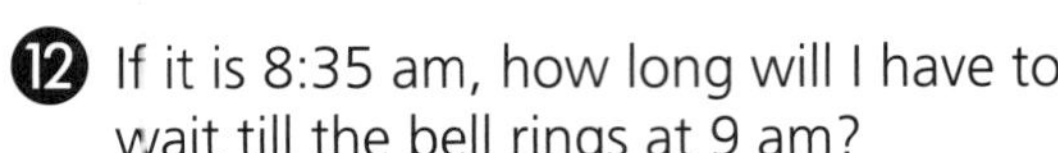 If it is 8:35 am, how long will I have to wait till the bell rings at 9 am? ______
13. 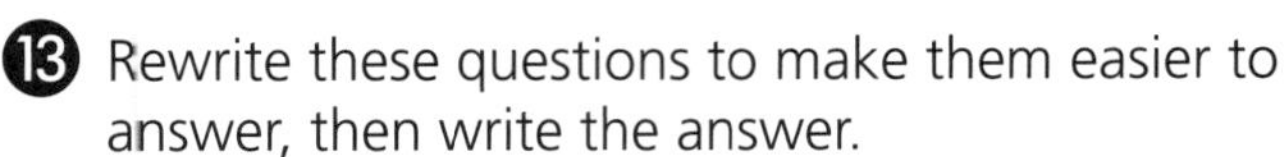 Rewrite these questions to make them easier to answer, then write the answer.
    a 163 + 298 = ______ + ______ = ______
      (−2) (+2)
    b 158 + 304 = ______ + ______ = ______
      (+4) (−4)
14. Today is Friday. What day will it be in 12 days time? ______
15. Minutes in:
    a three quarters of an hour ______
    b one and a half hours ______
    c 3 hours ______
16. What is the value of the 6 in 697 243? ______
17. a The fraction shaded is:

    ______ tenths and
    ______ hundredths.
    b The decimal for this model is: ______

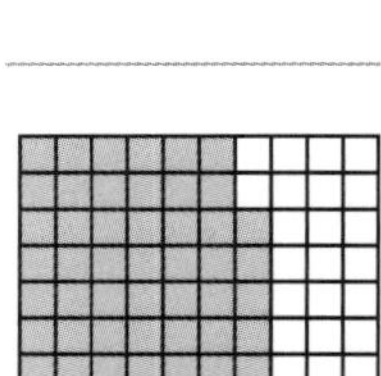

11:2 ☐ out of 17

1. 456 − 97 ______
2. 527 − 95 ______
3. 234 + 259 ______
4. 5346 + 35 ______
5.
   463
   + 298
6. 132 + 17 + 48 ______
7. 259 + 27 + 11 ______
8. 3405 − 2055 ______
9. 6278 + 24 ______
10.
   354
   + 578

11.  a Fraction shaded = $\frac{\square}{\square}$
    b Fraction not shaded = $\frac{\square}{\square}$

    c $\frac{3}{8} + \frac{\square}{\square}$ make 1 whole.

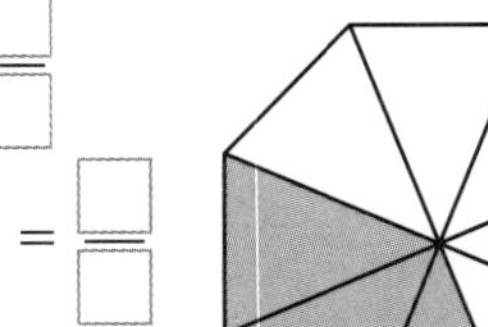

12.  Round 8354 to the nearest:

    a hundred ______ b thousand ______
13. My appointment is at 11 am. How long until my appointment if the time is 10:09 am? ______
14. I collected 435 cards and Rosie collected 259 cards. How many cards had we collected? ______
15. a 18 minutes after 9:25 am.
    ______
    b 30 minutes before 9:25 am.
    ______

16. Write in order from smallest to largest.
    $\frac{1}{2}$ $\frac{1}{8}$ $\frac{1}{4}$ ______
17. How many years in 1 century? ______

Roman numerals

1	I	one finger	
5	V	one hand	
10	X	two Vs	
50	L	half of a C	
100	C	centum = 100	

Write our numeral for:

a VIII ______ b XV ______ c LX ______

d CXX ______ e LXV ______ f CLX ______

Write the Roman numeral for:

g 22 ______ h 105 ______ i 212 ______

j 63 ______ k 75 ______ l 351 ______

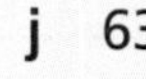

11:3   out of 9

1. Round 5376 to the nearest thousand. ______

2. **a** 3 years = ______ months
   b 40 years = ______ decades
   c 7 fortnights = ______ weeks

3. Complete the label for each clock.

morning

:

afternoon

:

4. Rewrite these questions to make them easier to answer, then write the answer.

   a 242 (−2, −20) + 478 (+2, +20) = ______ + ______ = ______

   b 576 (+4, −40) + 264 (−4, +40) = ______ + ______ = ______

5. I earned $385 this week and $286 last week. How much did I earn altogether? ______

6. Use am or pm to write:

   a 20:00 ______   **b** 07:00 ______

7. Match each fraction with a part of the decagon.

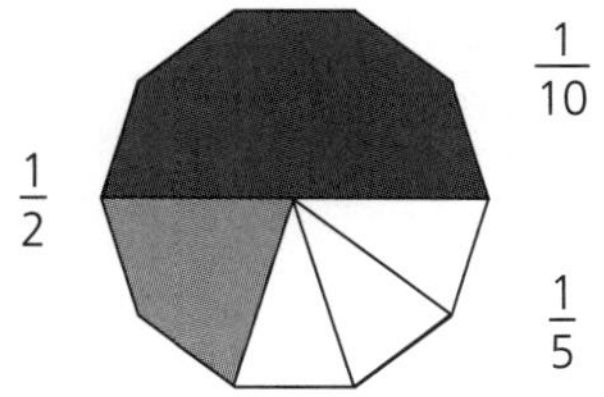

8. Circle the largest number:

   2 197 642   4 698 557   4 313 126

9. Colour $2\frac{1}{2}$ circles.

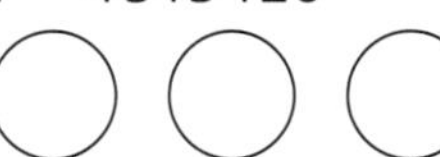

11:4   Extension   out of 7

1. If the time is 12:36, how long will it be until the end of school bell at 3 pm? ______

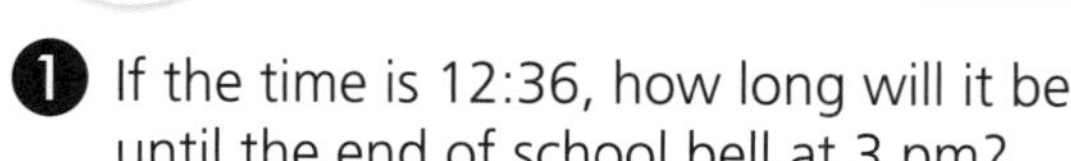

2. **a** 360 minutes = ______ hours
   b 264 weeks = ______ fortnights
   c 84 days = ______ weeks

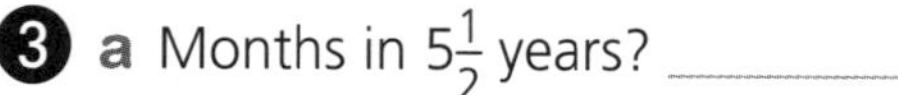

3. **a** Months in $5\frac{1}{2}$ years? ______
   b Days in $8\frac{3}{7}$ weeks? ______

4. 

   Every 4 minutes a dove flies away. How long would it take the doves to leave? ______

5. **a** Months in a decade? ______
   b Months in a century? ______

6. Lachlan, Andrew and David earned $254, $162 and $139 washing cars. How much did they earn altogether? ______

7. If the shaded part has the value given, find the value of the whole.

   a 

______

   b 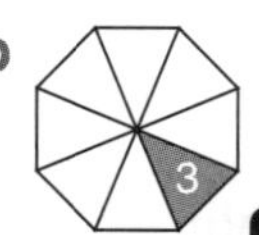

______

Challenge

Write questions that are equal to:

a 734 − 382	**b** 815 − 436	**c** 3291 − 2487
= ______	= ______	= ______
= ______	= ______	= ______
= ______	= ______	= ______
= ______	= ______	= ______
= ______	= ______	= ______

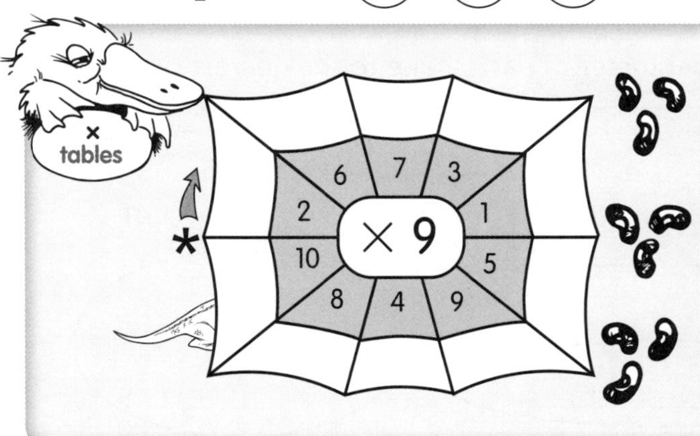

1 × 9		6 × 9	
2 × 9		7 × 9	
3 × 9		8 × 9	
4 × 9		9 × 9	
5 × 9		10 × 9	

12:1

☐ out of 17

1. 224 + 143 ____
2. 645 + 341 ____
3. 674 − 152 ____
4. 948 − 314 ____
5. $\begin{array}{r} 398 \\ +\,535 \\ \hline \end{array}$
6. 67 plus 9. ____
7. 82 take away 5. ____
8. 100 minus 31. ____
9. 435 + 99 ____
10. $\begin{array}{r} 767 \\ -\,435 \\ \hline \end{array}$

11. Write in order from smallest to largest.
    $\frac{1}{5}$ $\frac{1}{10}$ $\frac{1}{3}$ ____
12. **a** Write 6:00 pm in 24-hour time. ____
    b Write 10:00 pm in 24-hour time. ____
13. Rewrite these questions to make them easier to answer, then write the answer.
    a 195 (−4, −30) + 366 (+4, +30) = ____ + ____ = ____
    b 378 (−7, −10) + 483 (+7, +10) = ____ + ____ = ____
14. I bought 2 lengths of wood, each 187 cm long. What was the total length of wood? ____
15. **a** 5 minutes before 1:05? ____
    b 10 minutes before 1:05? ____

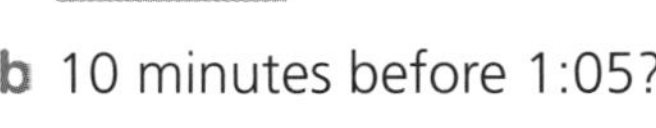

16. 0·42, 0·41, 0·40, ____, ____, ____
17. Write the digital time using am or pm.
    ____ : ____ ____

12:2

☐ out of 16

1. 378 − 98 ____
2. 346 − 97 ____
3. 326 + 258 ____
4. 6473 + 18 ____
5. $\begin{array}{r} 375 \\ +\,135 \\ \hline \end{array}$
6. 243 + 15 + 17 ____
7. 436 + 9 + 14 ____
8. 4207 − 2104 ____
9. 3025 − 1985 ____
10. $\begin{array}{r} 756 \\ -\,128 \\ \hline \end{array}$
11. I bought 2 oranges. One was 214 g and the other was 197 g.
    What was the total mass? ____
12. 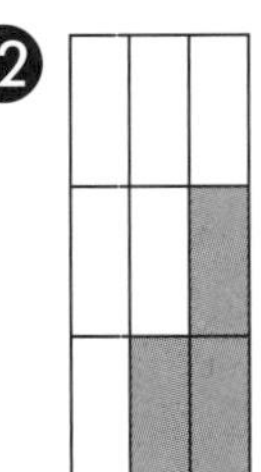
    a Fraction shaded = $\frac{\square}{\square}$
    b Fraction not shaded = $\frac{\square}{\square}$
    c $\frac{3}{9}$ + $\frac{\square}{9}$ make 1 whole.

13. **a** 584 (+5) − 395 (+5) = ____ − ____ = ____
    b 462 (+7, +20) − 173 (+7, +20) = ____ − ____ = ____
14. Write 13:00 using am or pm time. ____
15. My trip took 34 minutes. When did I arrive if I left home at 8:30 am? ____

16. Write the missing fraction and decimal.

34%	―	0·____

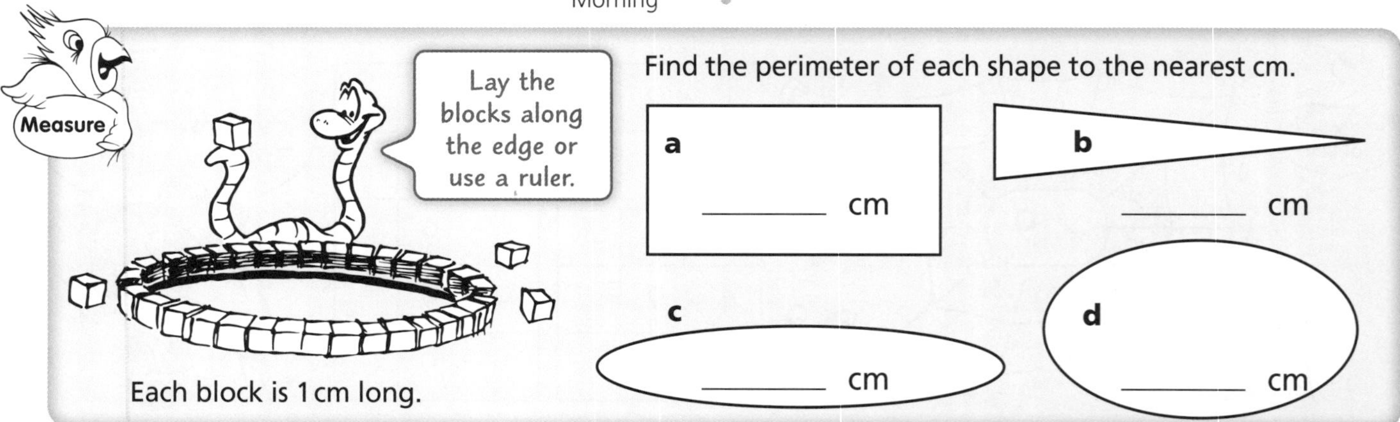

Find the perimeter of each shape to the nearest cm.

a ____ cm

b ____ cm

c ____ cm

d ____ cm

 ISBN 978 0 6557 0912 1

12:3 out of 9

1. $\frac{7}{10} + \frac{2}{10}$ ______

2. I bought 654 g of mushrooms and 357 g of carrots. What was the total mass? ______

3. Write 1800 using am or pm time. ______

4.

Bus timetable	
May St	08:54
Joy Ave	09:12
Red St	09:37
Fig St	09:48
Rae St	10:04

How long does it take to travel from:

a May St to Joy Ave? ______

b Red St to Rae St? ______

c May St to Rae St? ______

5. **a** Write the digital time using am or pm.

night

b Write this time using 24-hour time. ______

6. Mum swam 1 kilometre. Evan swam a quarter of this distance. How far did Evan swim? ______

7. Share 28 books among 4 children. How many each? ______

8. Write in order from largest to smallest.

8781344 8768367 8780033

______

9. 7·8, 7·9, 8·0, ______, ______, ______, ______

12:4 Extension out of 9

1. There was 1 L of water in a jug. I poured 236 mL into my glass and Kylie poured 564 mL into her glass. How much is left in the jug? ______

2. 8 kg of bananas cost me \$16.80. How much would 2 kg cost? ______

3. 47 + ☐ = 82

4. What fraction of a dollar is 10 cents?

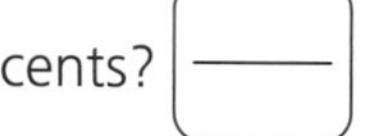

5. What is the time 6 hours after:

a 08:34? ______ **b** 14:29? ______

c 20:18? ______ **d** 23:37? ______

6. **a** Centuries in 2000 years. ______

b Decades in 700 years. ______

7. Four groups of $1\frac{1}{2}$. ______

8. **a** 938 + 402 ______ **b** 919 + 121 ______

9. **a** $5\frac{3}{4} - 2\frac{1}{4}$ ☐ **b** $7\frac{6}{8} - 5\frac{2}{8}$ ☐

Challenge

Write number sentences for addition of fractions, with a denominator of 10, e.g. $\frac{3}{10} + \frac{2}{10} = \frac{5}{10}$.

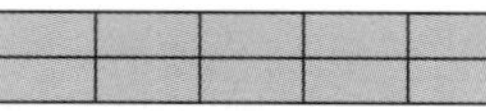

☆☆☆☆☆☆☆☆☆

÷,× tables

÷ 9: 63, 27, 72, 81, 45, 9, 36, 54, 90, 18

	4	1	9	0	7	5	2	11	6	3	8	10
× 4												

	8	4	7	1	11	0	5	10	2	3	6	9
× 7												

	3	0	4	8	5	10	7	1	6	11	2	9
× 9												

13:1

out of 17

1. 500 + 400 ______
2. 146 + 7 ______
3. 456 + 98 ______
4. 550 + 150 ______
5.
```
  576
+ 256
```
6. 74 − 30 ______
7. 40 shared by 5. ______
8. 7 times 4 mL. ______
9. Multiply 8 by 5. ______
10.
```
  693
- 274
```

11. Write 6 pm using 24-hour time. ______
12. Complete the first 7 multiples of 2.

    2, ______, ______, ______, ______, ______, ______
13. Circle the prime numbers. 3 9 12 7 4
14. Write the factors of 12.

    ______
15. Draw the top and side view of the pyramid above.

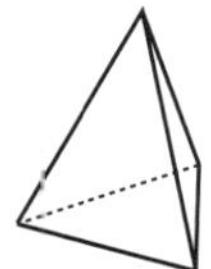

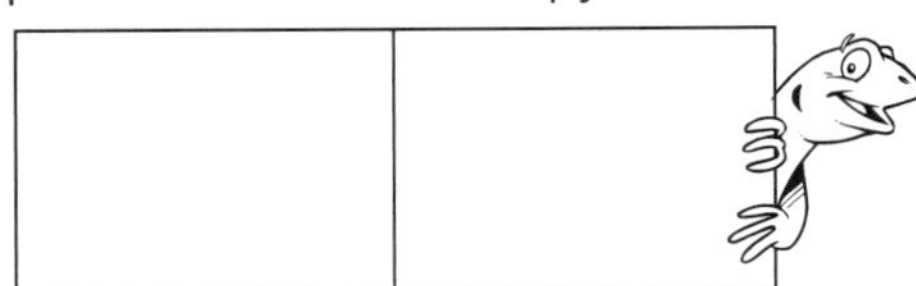

16. 

Write the total value of these coins. ______

17. Complete the pattern by shading.

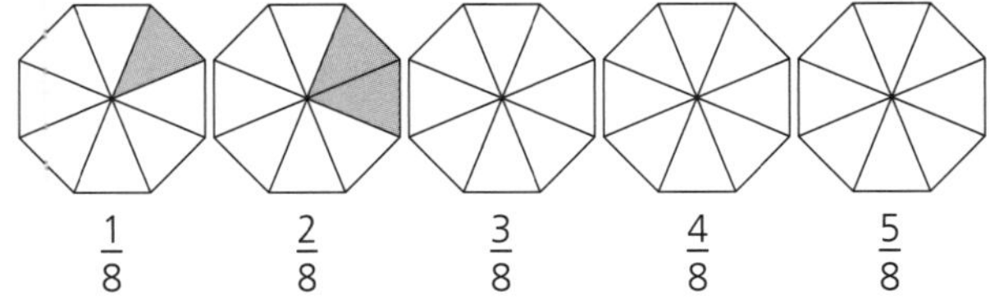

13:2

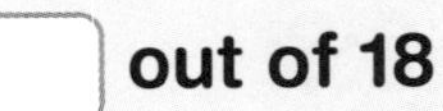

out of 18

1. 63 − 49 ______
2. 163 − 97 ______
3. 284 − 68 ______
4. 328 + 431 ______
5.
```
  673
+ 297
```
6. Multiply 8 and 9. ______
7. 56 shared by 7. ______
8. Divide 63 m by 9. ______
9. $\frac{1}{4}$ of 28. ______
10.
```
  896
- 248
```

11. Write 0400 using am or pm time. ______
12. **a** Write the digital time using am or pm.

    ______ : ______ ______

    b Write this time using 24-hour time. ______

13. Complete the first 7 multiples of 4 and 8.

    a 4, ______, ______, ______, ______, ______, ______

    b 8, ______, ______, ______, ______, ______, ______
14. A 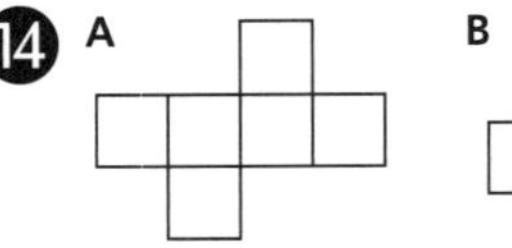 B 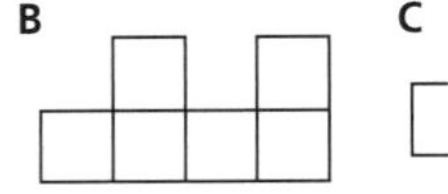 C 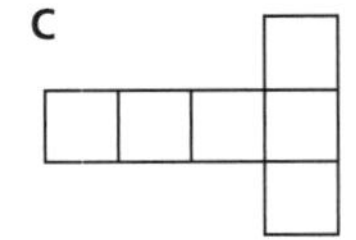

    Which of these nets will not make a cube? ______
15. Find the total if I bought a drink for \$3.55 and a fruit salad for \$5.68. ______
16. Write the factors of 18.

    ______
17. Round \$6.57 to the nearest 5 cents. ______
18. Circle the composite numbers. 16, 2, 8, 13, 1

Find the perimeter of each shape.

a Perimeter = ______ cm

b Perimeter = ______ cm

13:3 — out of 9

1 a 937 (+4) − 296 (+4) = ______ − ______ = ______

b 647 (+2, +30) − 268 (+2, +30) = ______ − ______ = ______

2 Convert 15:47 to 12-hour time. ______

3 Complete the first 7 multiples of 6.

6, ______, ______, ______, ______, ______, ______

4 Write the factors of 20.

____

5 Name this 3D object.

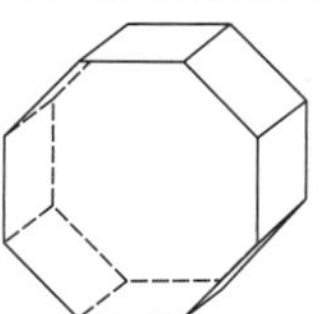

____

____

6

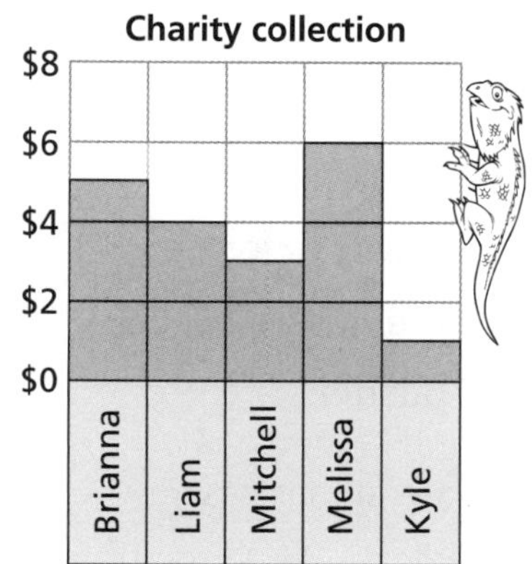

a How much money did Brianna give? ______

b What was the total amount of money given? ______

c How much more was needed to make a total of $30? ______

7 How many $50 notes make $300? ______

8 I started with $6.75 and spent $3.62. How much money do I have left? ______

9 Circle the prime numbers. 15 19 27 31

Extension

13:4 — out of 9

1 Write the factors of 100.

____

2 Seven cubes of side length 1 cm are glued together as shown. How many edges are on this object? ______

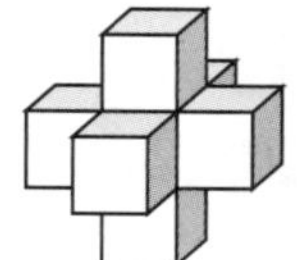

3 One quarter of $12.60. ______

4 One ice-cream sundae costs 55c. Find the cost of six.

______

5 How many 80c apples can I buy if I have $6? ______

6 What is my change from $100 if I bought a salad for $8.67 and drink for $4.57? ______

7 Find the smallest number that is a multiple of both 6 and 8. ______

8 After losing $20 and finding only $7, I had $78. How much did I start with? ______

9 If # means 'add 10' and ~ means 'subtract 3', then 7 # # # ~ ~ ~ = ______

Challenge

List multiples of:

a 2 ______________________

b 3 ______________________

c 4 ______________________

d 5 ______________________

e 6 ______________________

Turn to ID card D on page 9.

Give the answers for these numbers.

(4) ______ (5) ______ ______

(6) ______ ______ (7) ______ ______

(8) ______ ______ (9) ______ ______

(10) ______ ______ (12) ______

14:1 ☐ out of 18

1. $156 − $66 ______
2. $546 − $97 ______
3. Halve $68. ______
4. Double 56. ______
5. $\begin{array}{r} \$5.68 \\ +\ \$2.84 \\ \hline \end{array}$
6. $4 − $0.50 ______
7. 8 times 3 L. ______
8. 35 divided by 5. ______
9. 18 mL shared by 3. ______
10. $\begin{array}{r} \$9.74 \\ -\ \$5.84 \\ \hline \end{array}$
11. Complete the first 7 multiples of 3.
    3, ______, ______, ______, ______, ______, ______

12. This is a ______________________
    ______________________.
    It has ______ faces, and ______ edges and ______ corners.

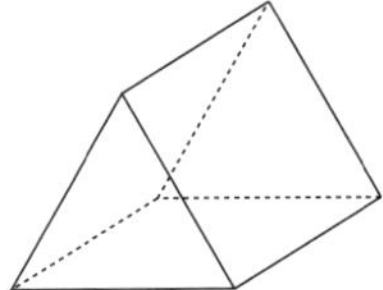

13. **a** Cents in $6.50. ______
    b 375 cents in decimal form. ______
14. Round $9.42 to the nearest 5 cents. ______
15. Write the factors of 28.
    ____
16. Is 28 a prime number? ______
17. Circle the polygons that have an acute angle.

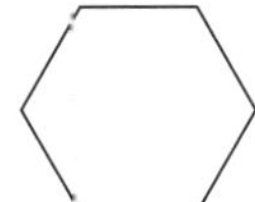  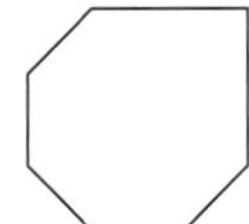 

18. Colour $2\frac{1}{2}$ tigers.

14:2 ☐ out of 17

1. 6 × 9 ______
2. 4 × 6 ______
3. 8 × 7 ______
4. 6 × 8 ______
5. $\begin{array}{r} \$8.13 \\ +\ \$2.27 \\ \hline \end{array}$
6. $167 − $38 ______
7. $465 + $383 ______
8. Multiply 7 by 30. ______
9. 63 g shared by 9. ______
10. $\begin{array}{r} \$6.83 \\ -\ \$3.97 \\ \hline \end{array}$

11. Write the factors of 32.
    ____
12. Circle the prime numbers. 29 33 57 19
13. Complete the first 7 multiples of 8.
    8, ______, ______, ______, ______, ______, ______
14. Write the mixed numeral for:
    a two and three tenths ______
    b six and three quarters ______
15. Name a quadrilateral that has:
    a two pairs of equal sides? ______
    b opposite sides parallel? ______
16. 

    Write the total value of these coins. ______
17. Draw an isosceles triangle on the left and a right-angled triangle on the right.

÷ tables

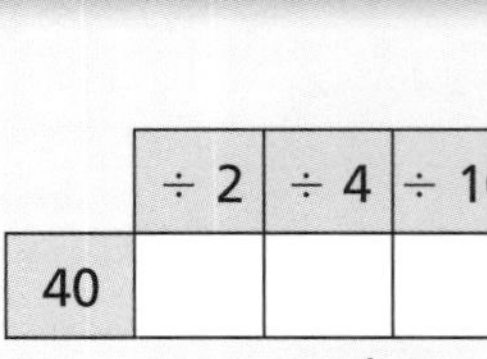

	÷ 7	÷ 35	÷ 5
35			

	÷ 2	÷ 4	÷ 10
40			

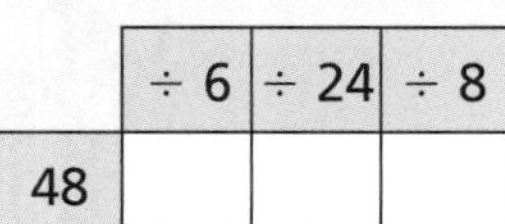

	÷ 6	÷ 24	÷ 8
48			

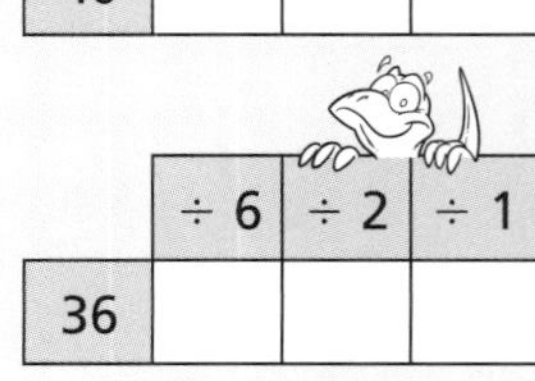

	÷ 6	÷ 2	÷ 1
36			

14:3 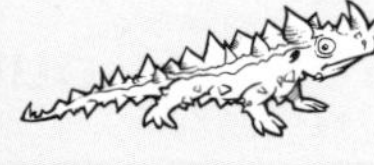 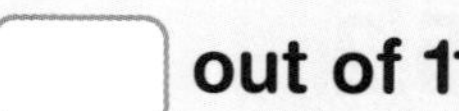 out of 11

❶ $6.73 − $1.45

❷ $3.85 − $2.37

❸ On a square pyramid, how many:

a faces? ______

b edges? ______

❹ Match words with the same meaning.

flip	rotation
slide	translation
turn	reflection

❺ Write the factors of 40.

____

❻ Write the mixed numeral for the shaded part.

a 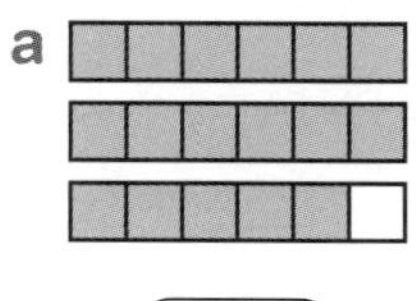

b 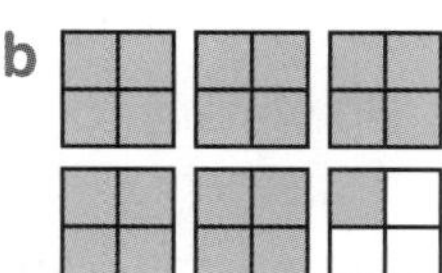

❼ Complete the first 7 multiples of 9.

9, ______, ______, ______, ______, ______, ______

❽ a Is a rectangle a special kite? ______

b Is a square a special rectangle? ______

c Is a rhombus a special kite? ______

❾ Write the numeral with 24 ones and 46 hundredths and 6 thousandths. ______

❿ The number before 7 536 097. ______

⓫ a $\frac{1}{5}$ of 20 = ______ b $\frac{1}{3}$ of 15 = ______

14:4 Extension out of 6

❶ 12 apples and 4 oranges cost $3.20. How much will 3 apples and 1 orange cost? ______

❷ If the fish facing right was reflected back and forth, which way would it be facing on the:

a 29th reflection? ______ b 106th reflection? ______

❸ 12 is a multiple of which counting numbers? ______

❹ If this pattern continues, how many hexagons would be in row:

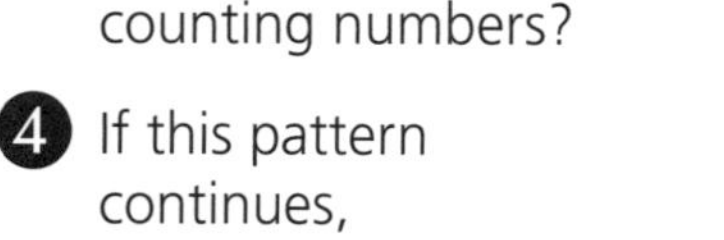

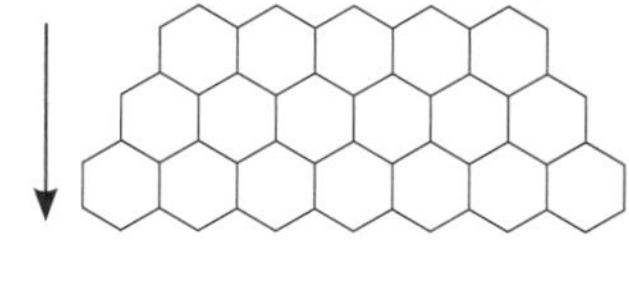

a 10? ______ b 100? ______

❺ 146, 171, 196, ______, ______, ______

❻ Shade $\frac{6}{12}$ of this shape.

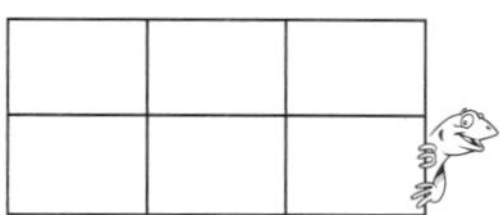

Challenge

Draw and label different types of triangles.

To round to the nearest 5 cents, give the closest answer that ends in 5 or 0.

Cash registers round off correct to the nearest 5 cents if you pay with cash.

a Is $19.98 closer to $19.95 or $20? ______

Round off each of these to the nearest 5 cents.

b $17.68 ______	c $21.01 ______	d $33.33 ______
e $18.97 ______	f $46.96 ______	g $52.49 ______

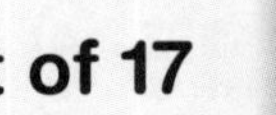

15:1

out of 17

1. 15 ÷ 3 ______
2. 16 ÷ 4 ______
3. 20 ÷ 5 ______
4. 18 ÷ 6 ______
5. $5.26 − $1.59
6. 6 × ______ = 30
7. 5 × ______ = 25
8. 3 × ______ = 12
9. 40 divided by 4. ______
10. $7.85 − $6.97

11. The value of the 4 in 9 348 738 is ______.
12. Does 3 hundredths equal 0·03? ______
13. How many groups of 5 stars?

Number of groups: ______

Number left over: ______

14. What is one share, if 28 toys are shared among:
    a 2 boxes? ______ b 4 boxes? ______
15. How many tens in 56?
    ______ and ______ left over.
16. 1200, 1300, ______, ______, ______
17. a Draw a quadrilateral that has opposite sides equal.
    b Draw an obtuse-angled triangle.

15:2

out of 17

1. 42 ÷ 6 ______
2. 49 ÷ 7 ______
3. 7 × ______ = 56
4. 9 × ______ = 36
5. $4.83 + $1.96
6. 36 shared by 4. ______
7. 54 divided by 9. ______
8. 46 ÷ 7 = ______ r ______
9. 53 ÷ 8 = ______ r ______
10. $8.57 − $3.93

11. Circle the scalene triangle.
    Cross the equilateral triangle.

 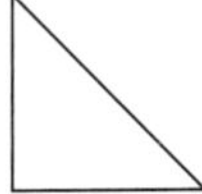 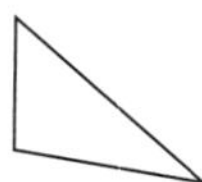

12. Weeks in 56 days. ______
13. Felicity planted plants in rows of 5.
    a How many rows could she plant if she had 37 plants? ______
    b How many plants would she have left over? ______
14. What solid can be made from this net?
    ______

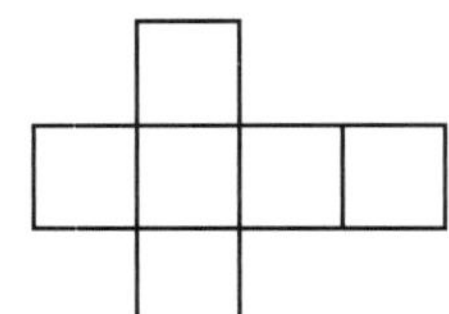

15. a 1 kg = ______ g b 1 L = ______ mL
16. 

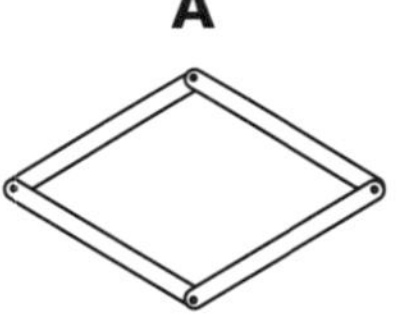

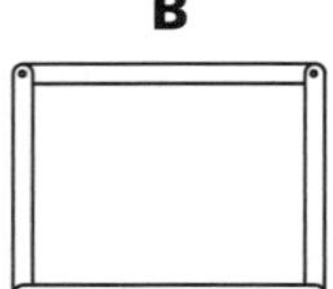

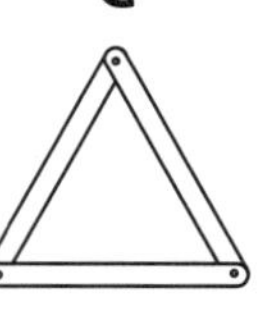

Which shape is rigid? ______

17. 53 = ______ × ______ + ______

Turn to ID card B on page 7.

Give the answers for these numbers.

(2)	______ lines	(3)	______ lines
(4)	______ line	(5)	______ line
(14)	______	(15)	______
(16)	______	(17)	______
(18)	axis of ______	(19)	______ of symmetry

15:3 out of 9

1. $9.28 − $4.63

2. $7.53 − $5.38

3. a $5\overline{)34}$ r b $8\overline{)77}$ r

4. Draw an obtuse-angled scalene triangle.

5. Share 19 helmets among 4 Vikings.
   One share = ______
   Left over = ______

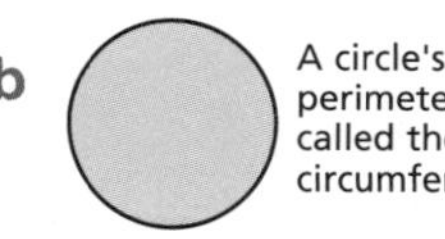

6. a 47 = ____ × ____ + ____
   b 33 = ____ × ____ + ____
   c 59 = ____ × ____ + ____

7.

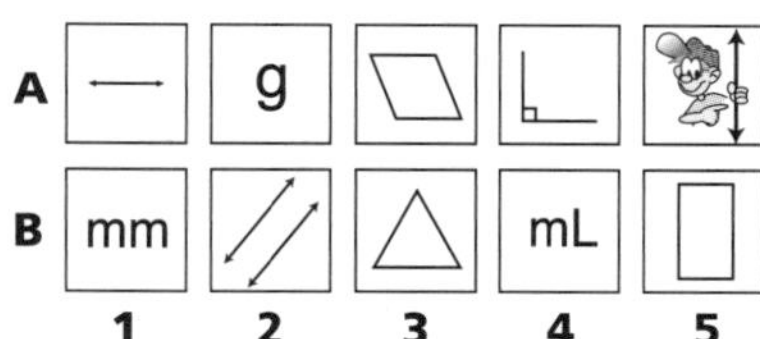

	1	2	3	4	5
A	—	g	parallelogram	right angle	
B	mm	parallel lines	triangle	mL	rectangle

a What is shown at 5B? ______

Write the coordinates for the:

b parallelogram ______

c parallel lines ______

8. a 3 km = ______ m b 4 m = ______ cm

9. Write in order from smallest to largest.

$\frac{1}{5}$ $\frac{1}{10}$ $\frac{1}{3}$ ______

15:4 out of 6

1. If we multiply together a number from each column, in how many ways can we get a product of 24? ______

	1	
2	2	2
8	3	3
4	4	4
	6	6

2. If 6 February was a Friday, what day was 28 January? ______

3. I earned 10 times more than I spent. I earned $970.
   a How much did I spend? ______
   b How much of this would I have left? ______

4. a ☐ ÷ 3 = 17   ☐ = ______
   b ☐ × 4 = 48   ☐ = ______

5. I bought 6 pens for $24. How much would it cost for:
   a one? ______   b two? ______

6. Estimate each perimeter to the nearest centimetre.
   a   b   A circle's perimeter is called the circumference.

Challenge

Write as much as you can about the number 6 534 108.

______

a

	2	4	6	8	10
× 2					

b

	2	4	6	8	10
× 5					

c

	1	2	3	4	5
× 4					

d

	1	2	3	4	5
× 10					

e

	6	3	8	7	9
× 0					

f

	4	7	3	9	6
× 1					

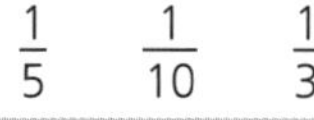

× tables

16:1   ☐ out of 17

1. 7×4 ____
2. $28 \div 4$ ____
3. 5×6 ____
4. $30 \div 5$ ____
5. $\begin{array}{r} 319 \\ -\ 155 \\ \hline \end{array}$
6. 8 divided by 4. ____
7. 12 shared by 3. ____
8. $5 \times$ ____ $= 20$
9. $4 \times$ ____ $= 24$
10. $\begin{array}{r} 925 \\ -\ 453 \\ \hline \end{array}$

11. Draw a right-angled triangle.
12. True or false? A rectangle is a special kite. ____
13. I have 28 plants. If I sell half of them today, how many will be left? ____
14. Four aeroplanes, all the same, have 32 lights altogether. How many lights are on each plane? ____

15. What object has position:
    a upper left?

    b lower right?

    c upper right?
    ____

16. Write the size of this angle.
    ____

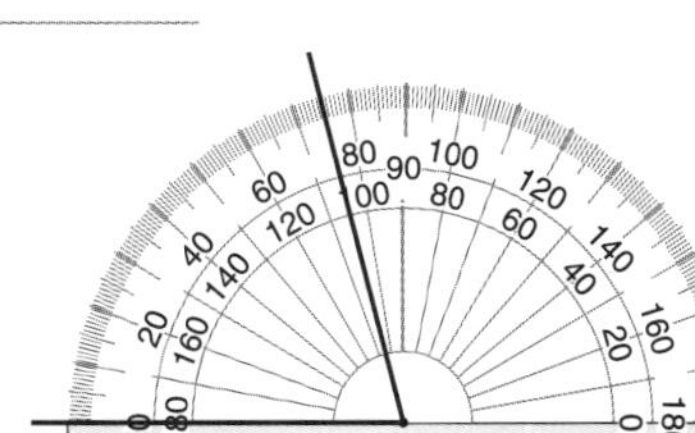

16:2   ☐ out of 15

1. 8×9 ____
2. 6×8 ____
3. $40 \div 8$ ____
4. $63 \div 7$ ____
5. $\begin{array}{r} 700 \\ -\ 365 \\ \hline \end{array}$
6. 58 plus 17. ____
7. Add 56 to 58. ____
8. 300 minus 43. ____
9. 72 divided by 8. ____
10. $\begin{array}{r} 800 \\ -\ 372 \\ \hline \end{array}$

11. 

    Find four equal shares of the birds.

    One share = ____ left over = ____

12. This is the net of a:
    s ____
    p ____

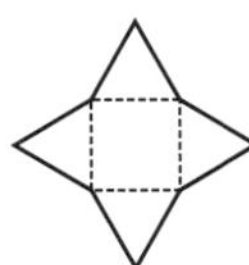

13. a $51 =$ ____ $\times$ ____ $+$ ____
    b $67 =$ ____ $\times$ ____ $+$ ____
    c $73 =$ ____ $\times$ ____ $+$ ____
14. John's team scored 435 goals this season. George's team scored 367. What was the difference between the number of goals scored for their teams? ____
15. How many right angles are the same size as one:
    a straight angle? ____
    b revolution? ____

Turn to ID card D on page 9.
Give the answers for these numbers.

(1) ____   (2) ____ or ____
(3) ____   (11) ____
(15) net of a ____   (16) net of a ____
(17) net of a ____   (18) net of a ____
(19) net of a ____

 • *AUSTRALIAN SIGNPOST MATHS NSW 5 MENTALS* • ISBN 978 0 6557 0912 1

16:3 ☐ out of 10

1. 743 − 586

2. 814 − 677

3. 674 − 297

4. $5\overline{)29}$ r

5. $7\overline{)67}$ r

6. 0·08, 0·09, 0·10, ______, ______, ______

7. These are shared among four collectors.

One share = ______

Left over = ______

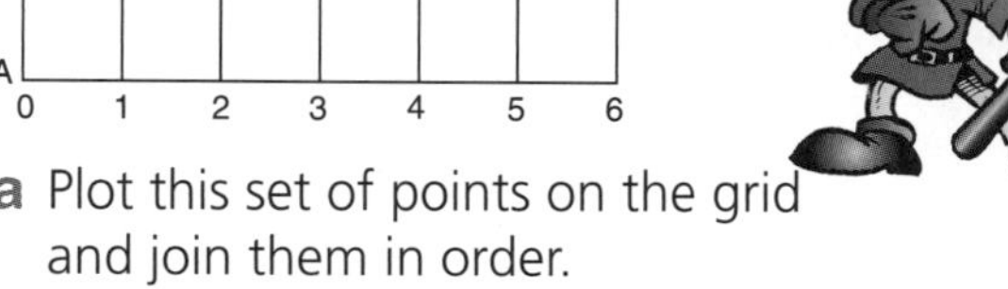

8. 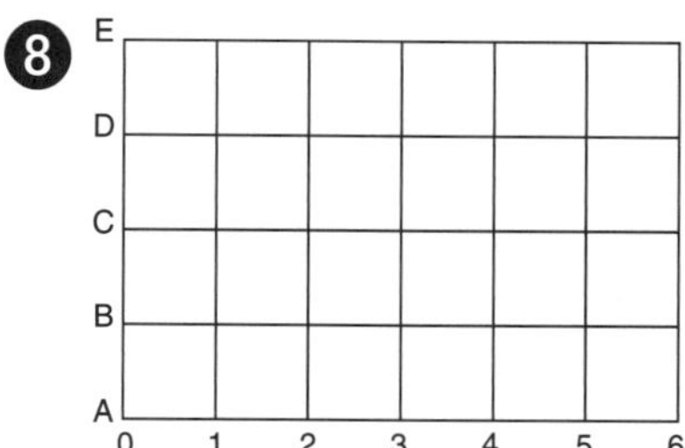

a Plot this set of points on the grid and join them in order.
1B, 2D, 3D, 4D, 5B, 4B, 3B, 2B, 1B

b What shape is formed? ______

9. a Colour 6 tenths red and 3 tenths blue.

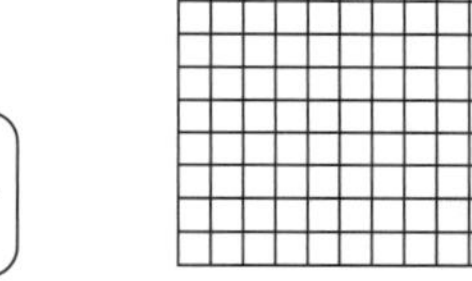

b What fraction would not be shaded? ☐

c Write this fraction as a decimal. ______

10. Write the size of this angle.

______

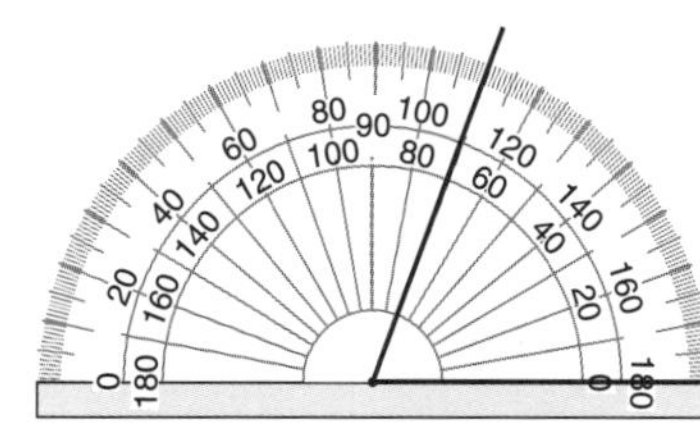

16:4 Extension ☐ out of 5

1. At the shops I spent \$356 and my sister spent \$275. How much less than \$800 did we spend? ______

2. 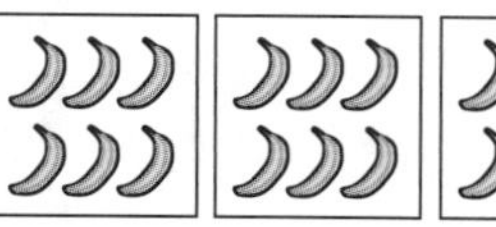

How many bananas in 15 groups of 6? ______

3. How many three-digit numbers can you make using the digits 2, 3 and 4 if:

a the digits cannot be used more than once in each number? ______

b the digits can be used more than once in each number? ______

4. 

How many matches are needed to make a line of 30 triangles? ______

5. $2 \times 20 \times 7$ ______

Challenge

Draw and label different types of angles.

Share 14 flowers among 3 people.
Answer: 4 each with 2 left over.

a Share 20 coins among 3 students.

①②③/①②○/○○○○
○○○○○○○○○○○

☐ coins each, with ☐ left over.

b Share 27 cards among 4 girls.

1	2	3	4	1	2			

☐ cards each, with ☐ left over.

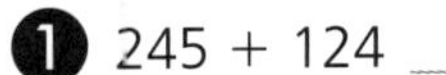

17:1 out of 17

1. 245 + 124 ______
2. 585 − 302 ______
3. 8 × 4 ______
4. 6 × 4 ______
5. $\begin{array}{r} 594 \\ -\,438 \\ \hline \end{array}$
6. 8 × ______ = 24
7. 35 divided by 5. ______
8. 20 shared by 4. ______
9. 35 + 63 − 63 ______
10. $\begin{array}{r} 7261 \\ +\,1573 \\ \hline \end{array}$
11. Write the size of this angle. ______

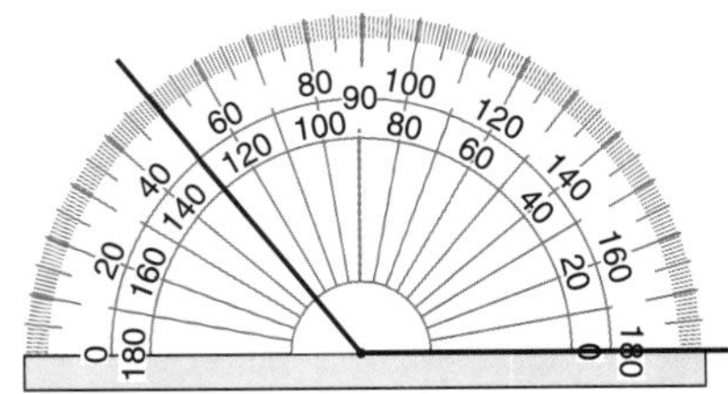

12. How many right angles in a square? ______
13. **Animals seen**

Dogs	卌 卌 卌 III	Cats	卌 卌 卌 II

   a How many dogs were seen? ______
   b How many cats were seen? ______

14. 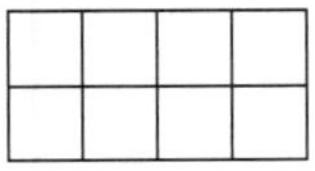 (grid of 8 squares)

   a $\frac{7}{8} - \frac{3}{8} =$ ______ b $\frac{4}{8} - \frac{2}{8}$ ______
   c $\frac{6}{8} - \frac{1}{8} =$ ______ d $\frac{6}{8} - \frac{5}{8}$ ______

15. 67·356 = ____ tens, ____ ones, ____ tenths, ____ hundredths and ____ thousandths
16. Circle the smaller decimal.
   a 0·8 0·17 b 0·1 0·09
17. Write $\frac{19}{5}$ as a mixed numeral. ______

17:2 out of 18

1. 8 × 8 ______
2. 6 × 8 ______
3. 9 × 6 ______
4. 45 ÷ 9 ______
5. $\begin{array}{r} 900 \\ -\,538 \\ \hline \end{array}$
6. 354 plus 298. ______
7. 567 minus 278. ______
8. 8 × ______ = 72
9. 4 × ______ = 36
10. $\begin{array}{r} 1681 \\ +\,4267 \\ \hline \end{array}$

11. What is left if I had $8 and spent $5.35? ______
12. Name the type of angle if its size is 67° ______
13. My family drank 4856 mL of water on Tuesday and 3645 mL on Wednesday. How much water did we drink altogether?
   ______ mL or ____ L ______ mL
14. Write the number:
   a 26 million ______
   b 35 billion ______
15. a Grams in $5\frac{1}{2}$ kg ______
   b Millimetres in $2\frac{1}{2}$ cm ______

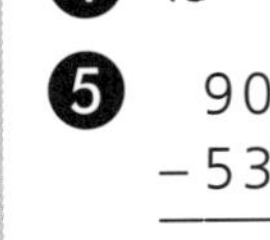

16. Try to use mental strategies to find:
   a 600 − 375 ______ b 300 − 287 ______
17. 82·293 = ____ tens, ____ ones, ____ tenths, ____ hundredths and ____ thousandths
18. a Draw a reflex angle. b Draw a right angle.

× tables

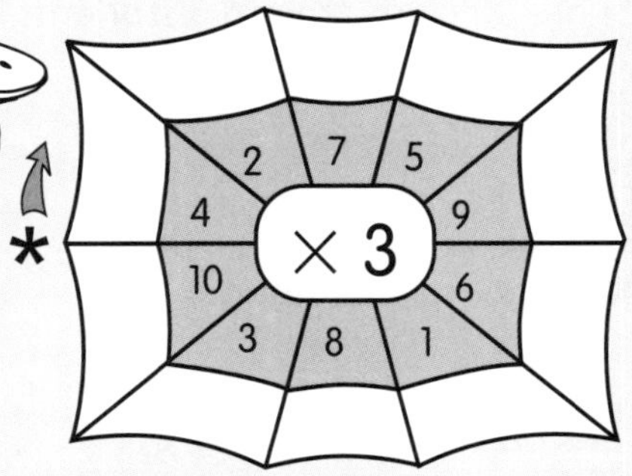

* × 3: 2, 7, 5, 9, 6, 1, 8, 3, 10, 4

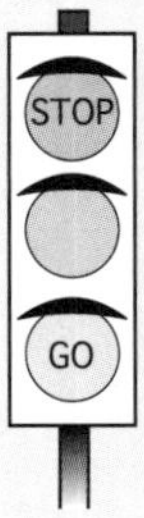

Time: ______

STOP GO

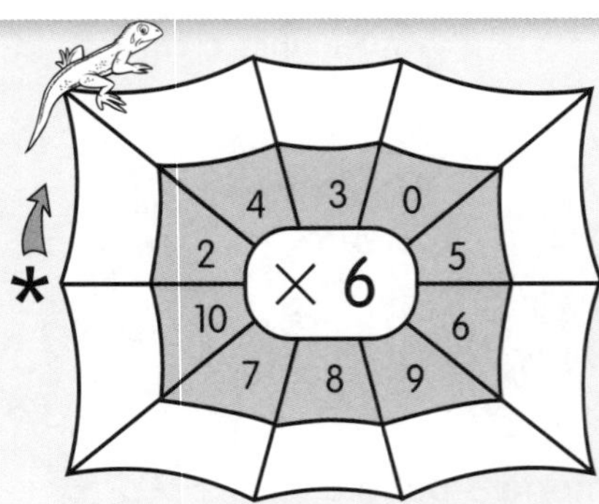

* × 6: 4, 3, 0, 5, 6, 9, 8, 7, 10, 2

Time: ______

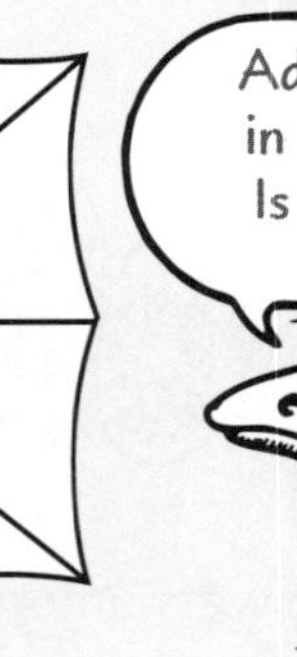

 ISBN 978 0 6557 0912 1

17:3 ☐ out of 12

1. $397 - 213$

2. $635 - 432$

3. $842 - 735$

4. $7\overline{)26}$ r

5. $9\overline{)76}$ r

6. $4452 + 4749$

7. $6235 + 2797$

8. What is left if I had $6 and spent $3.76? ______

9. Write the size of this angle. ______

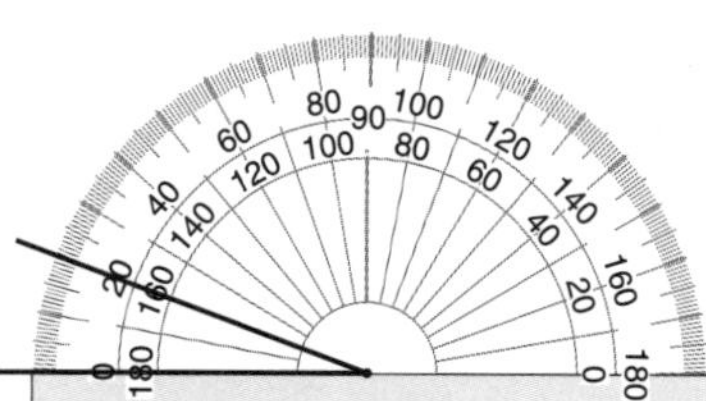

10. Susanna's suitcase weighed 4527 g and Abigail's weighed 3914 g. How much did they weigh altogether?

______ g **or** ______ kg ______ g

11.

1–10	IIII
11–20	𝍸 𝍸 II
21–30	𝍸 𝍸 III

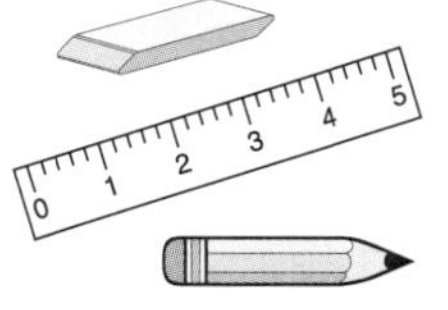

a How many students scored 11–20 marks? ______

b Do we know how many students scored 30? ______

c How many scored over 10? ______

12. What is the value of the 9 in 45·397? ______

17:4 Extension ☐ out of 6

1. We raised $5366 at our first fundraiser and $2589 at our second. How much more do we need to reach our target of $9000? ______

2. How many angles, out of A to H, are:

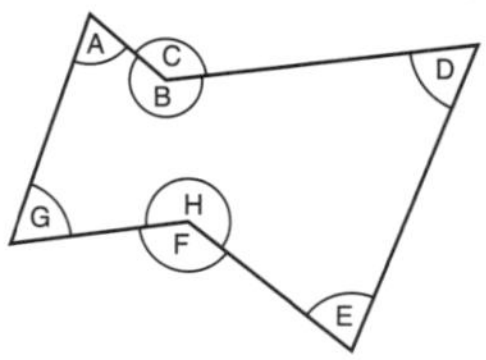

a acute? ______

b reflex? ______

3. The whole has a value of 20. Write the value of each shaded part.

20

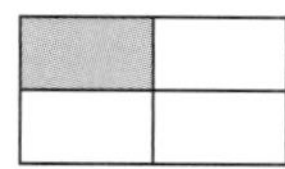

______ ______

4. Six equal angles make up a straight angle. How big is each angle? ______

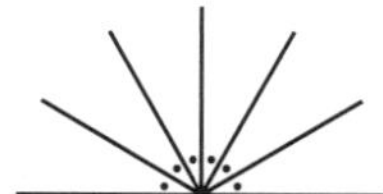

5. Complete the pattern.
1, 3, 7, 15, ______, ______, ______, ______

6. If $24 \times 32 = 768$, then $768 \div 24 =$ ______

Challenge

Write everything you know about the number 83·561.

______

______

______

______

______

18:1 out of 19

1. 6 × 4 ______
2. 15 ÷ 3 ______
3. 21 ÷ 7 ______
4. 4 × 8 ______
5. $\begin{array}{r} 5368 \\ +\,2581 \\ \hline \end{array}$
6. 560 plus 29. ______
7. Add 299 to 463. ______
8. 325 minus 30. ______
9. 5 × ______ = 45
10. $\begin{array}{r} 4539 \\ -\,2546 \\ \hline \end{array}$
11. Is an obtuse angle greater than an acute angle? ______
12. Fill in the missing words.

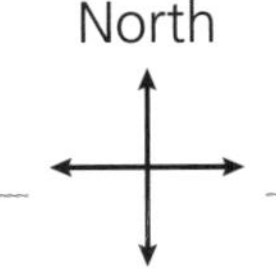

13. Make a tally for the letters E, S, U, F and M from these children's names.
Jessica, Shaun, Mary, Geoffrey, Kyle

E	S	U	F	M

14. Circle the smallest decimal.
0·14 0·43 2·1 1·09
15. Write 67 hundredths as a decimal. ______
16. Write 0·52 as a fraction.

17. Is 5 a factor of 35? ______
18. The value of the 3 in 45·937? ______
19. **a** Draw an acute angle **b** Draw an obtuse angle

18:2 out of 14

1. 30 ÷ 5 ______
2. 7 × 7 ______
3. 465 + 497 ______
4. 9 × 6 ______
5. $\begin{array}{r} \$24.75 \\ +\,\$59.26 \\ \hline \end{array}$
6. 6 × ______ = 36
7. 9 × ______ = 63
8. 58 more than 385. ______
9. Double 263. ______
10. $\begin{array}{r} 5463 \\ -\,3656 \\ \hline \end{array}$
11. Which angle is:
a an acute angle? ______
b an obtuse angle? ______

12. Write the type of angle if its size is:
a 62° ______
b 177° ______
13. Use the table in 18:1, Question 13 to complete this graph. Which letter was used:
a most? ______
b 3 times? ______

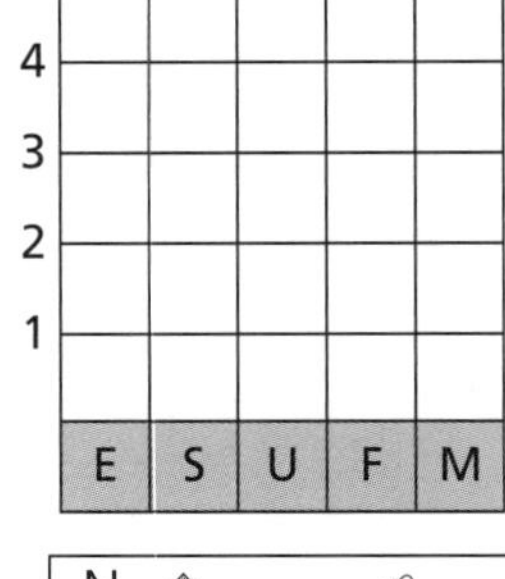

14. On the map, Kristen is standing at **X**. What does she see if she looks:

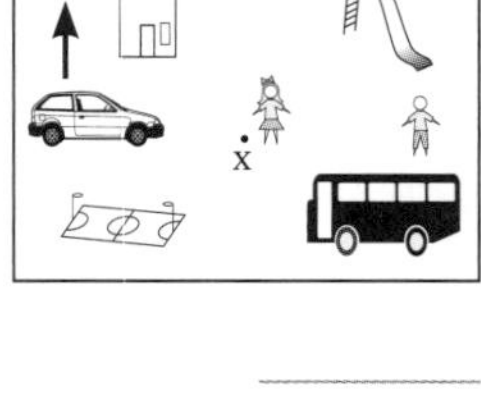

a west? ______
b east? ______
c south-east? ______
d north-west? ______

Concept

Compass points

The direction halfway between north and east is called north-east.
What is the direction halfway between:

a south and east? ______
b south and west? ______
c north and west? ______

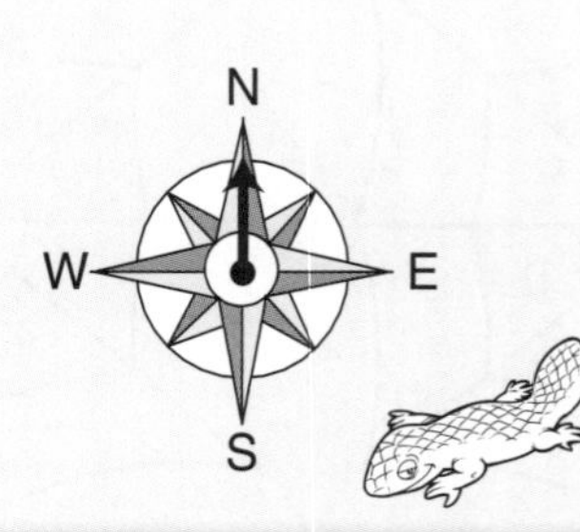

Answers

ID card answers

ID card A

1 millimetres 2 centimetres 3 metres 4 kilometres 5 millilitres 6 litres 7 grams 8 kilograms 9 seconds 10 minutes 11 hours 12 square centimetres 13 square metres 14 metre 15 1 hectare 16 hectares 17 square kilometres 18 1 cubic centimetre 19 degrees Celsius 20 5 × 7 21 4 × 10 22 20 ÷ 5 23 42 ÷ 6 24 even numbers 25 odd numbers 26 digits 27 1, 8, 2 and 4 28 7 29 5 30 remainder

ID card B

1 line 2 parallel lines 3 perpendicular lines 4 horizontal line 5 vertical line 6 corner or vertex 7 arm of an angle 8 acute angle 9 right angle 10 obtuse angle 11 straight angle 12 reflex angle 13 revolution 14 tessellation 15 flip 16 slide 17 turn 18 axis of symmetry 19 axes of symmetry 20 number line 21 north, east, south, west 22 north-east, south-east, south-west 23 coordinates 24 tally 25 picture graph 26 dot plot 27 column graph 28 line graph 29 sector or pie graph 30 divided bar graph

ID card C

1 oval 2 triangle 3 square 4 rectangle 5 rhombus 6 trapezium 7 parallelogram 8 kite 9 quadrilaterals 10 pentagons 11 hexagons 12 octagon 13 regular shapes 14 irregular shapes 15 diagonals 16 decimal point 17 5 tens, 4 ones, 3 tenths, 9 hundredths 18 5 sixths shaded 19 3 out of 8 20 $\frac{3}{10}$ shaded 21 3 tenths shaded 22 0·3 shaded 23 $\frac{24}{100}$ shaded 24 0·36 shaded 25 3 tenths and 6 hundredths 26 19 out of 100 27 62% shaded 28 1·36 29 $1\frac{5}{10}$ or $1\frac{1}{2}$ 30 $0{\cdot}3 = \frac{3}{10}$, $0{\cdot}30 = \frac{3}{10}$

ID card D

1 face 2 vertex or corner 3 edge 4 cube 5 rectangular prism 6 triangular prism 7 hexagonal prism 8 triangular pyramid 9 square pyramid 10 rectangular pyramid 11 base 12 cylinder 13 cone 14 sphere 15 net of a cube 16 net of a square pyramid 17 net of a cylinder 18 net of a cone 19 net of a triangular prism 20 3 21 1 22 2 23 before noon 24 after noon 25 19:30 26 5:40 am 27 protractor 28 thermometer 29 compass 30 tape measure

1:1

❶ 50 ❷ 10 ❸ 7 ❹ 2 ❺ 23 ❻ 5 ❼ 500 ❽ 85 ❾ 32

❿ 777 ⓫ a 6 b 5 ⓬

⓭ a 32 b 25 ⓮ $16.50 ⓯ 3 ones, 2 tenths and 4 hundredths

⓰ 15 ⓱ a 600 b 900 c 700 ⓲ 10 ⓳ 7542

1:2

❶ 360 ❷ 4 ❸ 4 ❹ 6 ❺ 925 ❻ 10 ❼ 7 ❽ 3 ❾ 30

❿ 216 ⓫ 3 × 28 = 84, break down the question: (3 × 10) + (3 × 10) + (3 × 8). ⓬ 9 cm², 4·5 cm² ⓭ a 3 kg b 5 kg

⓮ You can put people into groups of 2, 3, 6, 7, 14, or 21.

⓯ 13 ⓰ 64

Activity

10, 12, 6, 16, 2, 18, 14, 8, 4, 20

36, 12, 42, 24, 60, 6, 42, 12, 54, 30

1:3

❶ 10 out of 100 or 1 out of 10 ❷ 4 ❸ 3622 ❹ 11

❺ 34, 54, 74, 94, 114 ❻ 35

❼ a 94 b 44

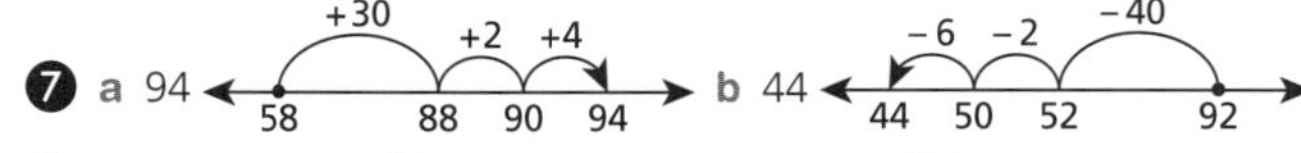

❽ a 133 b 243 ❾ 78 − 36 = 48 − 6 = 42 ❿ 12, 15, 18, 21, 24

1:4

❶ 383

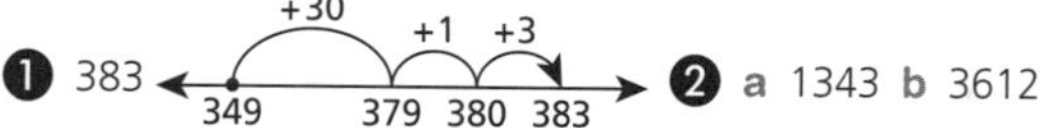

❷ a 1343 b 3612

❸ 16 L ❹ 70 ❺ 8 m 98 cm or 898 cm ❻ 20 ❼ 8 ❽ 10

Challenge

Answers will vary.

E.g.

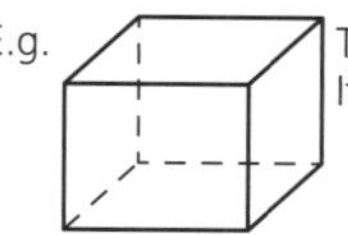

This is a rectangular prism.
It has 6 faces, 12 edges, and 8 vertices.

Activity

Answers will vary.

2:1

❶ 21 ❷ 24 ❸ 48 ❹ 27 ❺ 12 ❻ 20 ❼ 40 ❽ 10 ❾ 5

❿ 884 ⓫ 42 ⓬ a 12, 16 b yes

⓭ 0·7 ⓮ yes ⓯ a 7 b 6 c 7 d 6 ⓰ 5646

⓱ 456 mL ⓲ $4.15, $5.85 + 5c + 10c + $2 + $2, $5.85 → $5.90 → $6.00 → $8.00 → $10.00 ⓳

2:2

❶ 7000 ❷ 530 ❸ 40 ❹ 180 ❺ 225 ❻ 8 ❼ 9 ❽ 7

❾ $30.60 ❿ 999 ⓫ 13 × 3 = 39, break down the question into (10 × 3) + (3 × 3). ⓬ a 527 + 100 − 4 = 623 b 374 + 100 − 2 = 472 c 734 + 100 − 3 = 831 ⓭ $80 ⓮

⓯

	Thousands	Hundreds	Tens	Units
a	57	3	4	2
b	78	5	4	6
c	32	2	5	4

Activity

10, 50, 30, 100, 70, 40, 80, 20, 90, 60

5, 25, 15, 50, 35, 20, 40, 10, 45, 30

2:3

❶ 57 359 ❷ 956 941 ❸ 24 ❹ a 3 b 10

❺

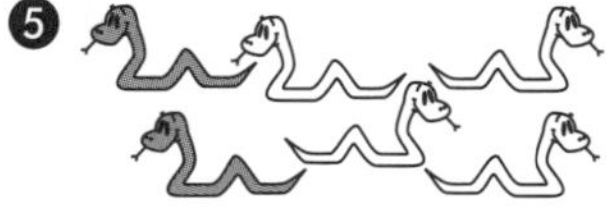

❻ a 3 h b 3 h ❼ a 57 hundredths b 43 out of 100

❽ $0.75 ❾ 62 ❿ 36 470

2:4

❶ a 100 b 75 c 125 ❷ 28 ❸ $22 ❹ $4.75 ❺ 17

❻ a 88 b 112 ❼ 540 ❽ 30 ❾ 5 ❿ 11

Challenge
Answers may vary, e.g. It is a decimal. It has 5 hundreds, 3 tens, 4 units, and 6 tenths. It becomes 500 when rounded to the nearest hundred.

Activity
27, 21, 18, 24, 3, 9, 12, 30, 15, 6
16, 36, 28, 8, 4, 40, 32, 24, 12, 20

3:1

1 21 **2** 51 **3** 9 **4** 800 **5** 792 **6** 14 **7** 8 **8** 47 **9** 40 **10** 25 **11** $1.30 **12** 28 **13** no **14** **a** 15 min **b** 25 min **15** **a** metres (m) **b** centimetres (cm) **16** **a** 35 + 40 = 75 **b** 24 + 50 = 74 **17** 5750·35

3:2

1 13 **2** 8 **3** 8 **4** 4 **5** 15 **6** 5 **7** 10 **8** 26 **9** 39 **10** 1176 **11** 120 **12** 42 **13** 58 **14** 18 **15** yes **16** more **17** **a** 6 **b** 12 **18**

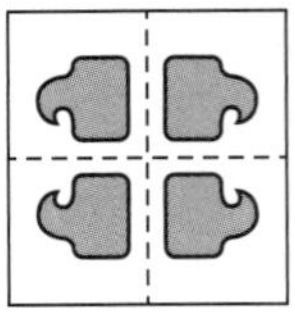

19 triangular pyramid

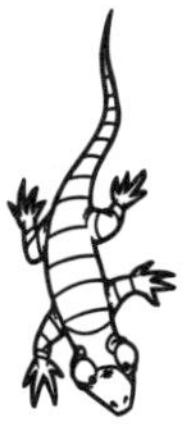

Activity
30, 6, 18, 27, 3, 12, 21, 9, 15, 24
6, 54, 36, 18, 42, 48, 24, 30, 12, 60

3:3

1 6 **2** 8 **3** 3 **4** July and August or December and January. **5** $13 **6** no **7** $2 or 200c **8** four hundred and seventy-eight thousand, three hundred and ninety-one. **9** A – acute, B – right, C – straight, D – reflex **10** **a** 600 **b** 0·09 **11** 6·38 **12** 734 099 is ticked, 734 902 is circled

3:4

1 48 **2** 64 **3** 5 **4** **a** $1.15 **b** 5c **5** 13 **6** 320 **7** 28

Challenge
Answers will vary. See 3:4, Question 7 for an example.

Activity
14, 42, 7, 28, 70, 49, 21, 63, 35, 56
35, 21, 49, 63
An odd number times an odd number gives an odd number.

4:1

1 12 **2** 15 **3** 4 **4** 8 **5** 210 **6** 2 **7** 2 **8** 8 **9** 5 **10** 426 **11** 12 **12** 6, 9, 12, 15, 18 **13** 845 320 **14** $24 **15** **a** 53 205 **b** 20 381 **16** 12 cm **17** 12 **18** $1.95 **19** 9 000 000 **20** 34

4:2

1 45 **2** 24 **3** 32 **4** 36 **5** 105 **6** 7 **7** 6 **8** 7 **9** 22 **10** 368 **11** 28 **12** $1.50 or 150c **13** 35 738 943, 35 729 499, 35 709 946 **14** **a** 18 426 364 **b** 4 200 000 **c** 23 000 000 000 **15** **a** 4647 **b** 14 980 **16** 865 450 034 **17** **a** T **b** T **c** F **d** T **18** $1.20

Activity
56, 16, 72, 32, 48, 40, 8, 64, 24, 80 **a** 56 **b** 48

4:3

1 89 104 500, 79 092 736, 79 026 748 **2** **a** 525 000 **b** 339 354 **3** 10, 15, 20, 25, 30 **4** $36 **5** **a** 78 475 043 **b** 8 074 005 **c** 78 000 000 017 **6** **a** 160c **b** $1.60 **7** 1924 **8** 4 000 000 **9** **a** $2\frac{1}{3}$ **b** $3\frac{2}{5}$

4:4

1 **a** 128 **b** 160 **2** 70 **3** 272 **4** **a** 48 **b** 53 **5** **a** 1 675 000 **b** 1 120 354 **6** 9 weeks

Challenge
Answers will vary. **a** 2, 4, 6, 8, 10, 12, 14, etc. **b** 3, 6, 9, 12, 15, 18, 21, etc. **c** 4, 8, 12, 16, 20, 24, 28, etc. **d** 5, 10, 15, 20, 25, 30, 35, etc. **e** 6, 12, 18, 24, 30, 36, 42, etc.

Activity
18, 54, 63, 27, 9, 45, 81, 36, 72, 90
9, 18, 27, 36, 45, 54, 63, 72, 81, 90

5:1

1 6 **2** 10 **3** 3 **4** 7 **5** 8 **6** 21 **7** 8 **8** 30 **9** 405 836, 405 657, 405 299 **10** 6 023 403 **11** **a** 6 **b** 6 **12** 23 512 **13** 16, 24, 32, 40, 48 **14** $\frac{1}{2}$ **15** $\frac{3}{5}$ **16**

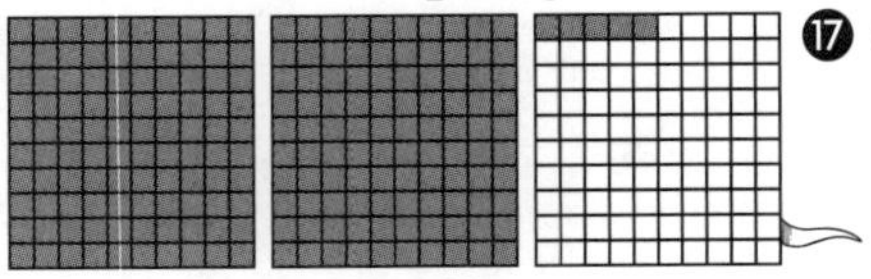

17 6·56 **18** 1·2, 1·4, 1·6, 1·8 **19** 4 **20** 5 r 1

5:2

1 42 **2** 54 **3** 8 **4** 8 **5** 27 **6** 63 **7** 35 **8** 10 **9** 49 **10** 3 **11** 6 r 2 **12** 7·2, 7·4, 7·6, 7·8 **13** 85 809 375, 85 700 374, 85 078 836 **14** **a** 8 **b** 8 **15** 8, 12, 16, 20, 24 **16** $\frac{3}{4}$ **17** $\frac{3}{10}$ **18** **a**

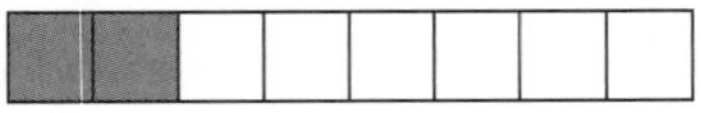

b 2 **19** $3\frac{9}{100}$ **20** 36 **21** **a**

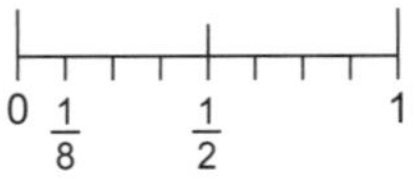

b $\frac{1}{2}$ will be circled.

Activity
4, 10, 6, 2, 8, 7, 1, 3, 9, 5
2, 5, 3, 1, 4, 9, 7, 10, 8, 6

5:3

1 $\frac{1}{10}, \frac{1}{5}, \frac{1}{4}$ **2**

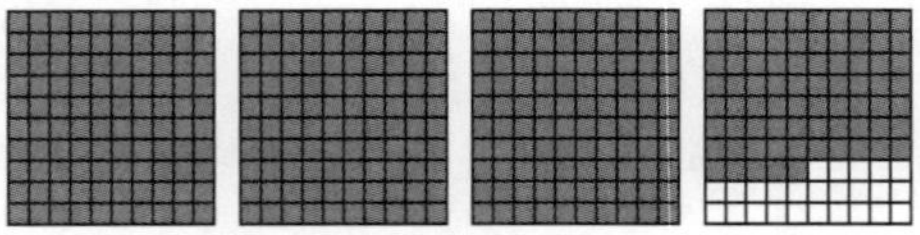

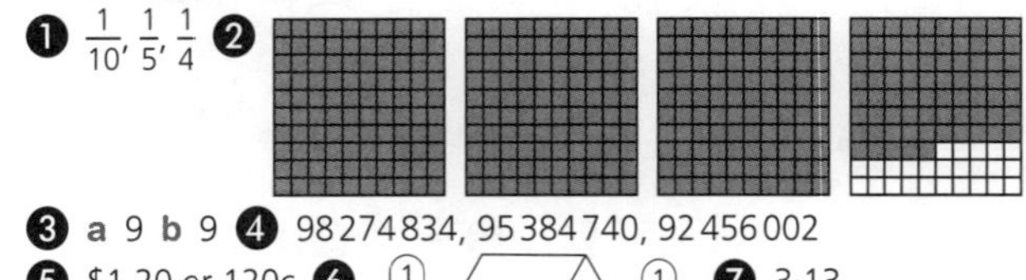

3 **a** 9 **b** 9 **4** 98 274 834, 95 384 740, 92 456 002 **5** $1.20 or 120c **6**

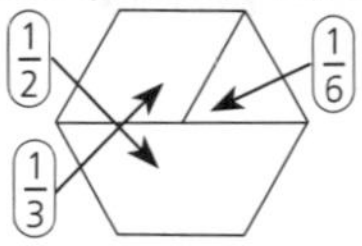

7 3·13 **8** **a** T **b** T **c** T **d** F **9** + **10** $2.85 or 285c **11** cm² **12** 6 r 5

© PEARSON AUSTRALIA 2024 • *AUSTRALIAN SIGNPOST MATHS NSW 5 MENTALS* • ISBN 978 0 6557 0912 1

5:4

❶ 60 ❷ a 40 b 33 ❸ 12 weeks ❹ 98 ❺ 30 ❻ 4, 16

Challenge

Answers will vary.

E.g. 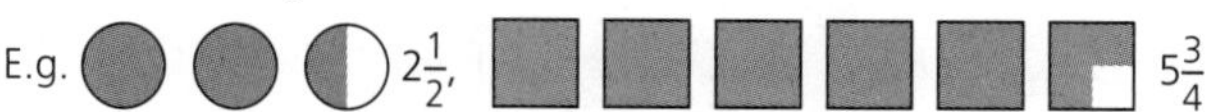 $2\frac{1}{2}$, $5\frac{3}{4}$

Activity

1, 8, 5, 2, 7, 9, 3, 10, 6, 4

5, 1, 10, 6, 2, 8, 4, 7, 3, 9

$8 \times 3 = 24$

$8 \times 6 = 48$

6:1

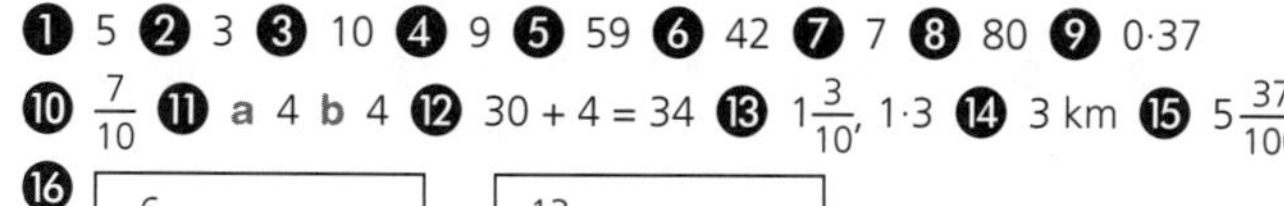

❶ 5 ❷ 3 ❸ 10 ❹ 9 ❺ 59 ❻ 42 ❼ 7 ❽ 80 ❾ 0·37

❿ $\frac{7}{10}$ ⓫ a 4 b 4 ⓬ 30 + 4 = 34 ⓭ $1\frac{3}{10}$, 1·3 ⓮ 3 km ⓯ $5\frac{37}{100}$

⓰ 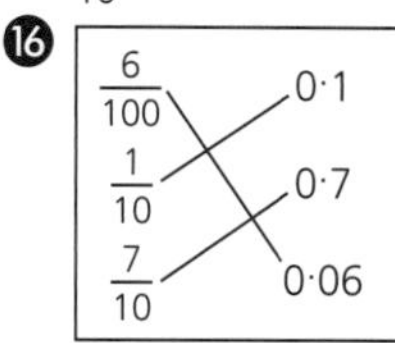 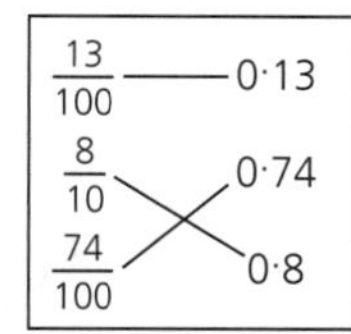

⓱ 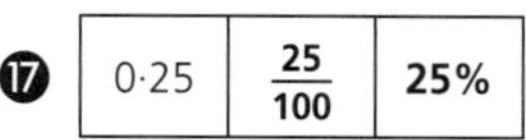

0·25	**$\frac{25}{100}$**	**25%**

0·50	**$\frac{50}{100}$**	**50%**

⓲ a 2 b 10 ⓳ 5 r 2

6:2

❶ 9 ❷ 7 ❸ 8 ❹ 9 ❺ 16 ❻ 10 ❼ 8000 ❽ 50

❾ \$100 + \$100 + \$20 + \$10 + \$1 + 20 c + 5 c ❿ $\frac{2}{3}$ ⓫ 64

⓬ $9\frac{94}{100}$ ⓭ a 719 000 b 857 354

⓮ a b 4 c $\frac{3}{4}$ or $\frac{12}{16}$

⓯ a 0, $\frac{1}{4}$, $\frac{1}{2}$, 1 b $\frac{1}{2}$ will be circled. ⓰ 6000 m

⓱ 0·8 ⓲ 6·6, 6·7, 6·8, 6·9 ⓳ 4·29 ⓴ 6 r 3

Activity

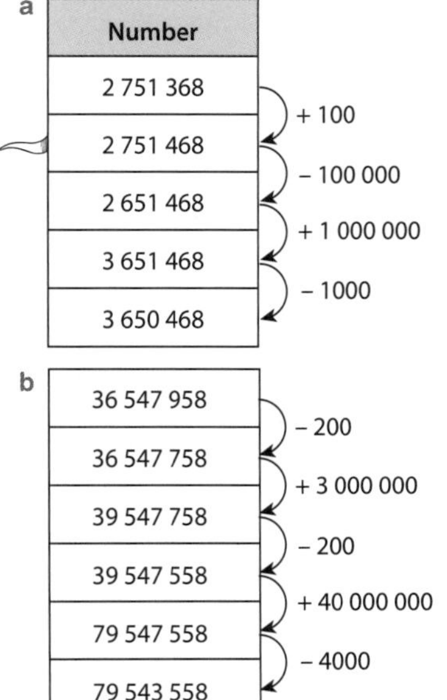

a

Number	
2 751 368	+ 100
2 751 468	− 100 000
2 651 468	+ 1 000 000
3 651 468	− 1000
3 650 468	

b

36 547 958	− 200
36 547 758	+ 3 000 000
39 547 758	− 200
39 547 558	+ 40 000 000
79 547 558	− 4000
79 543 558	

6:3

❶ 6 r 2 ❷ $\frac{1}{5}$, 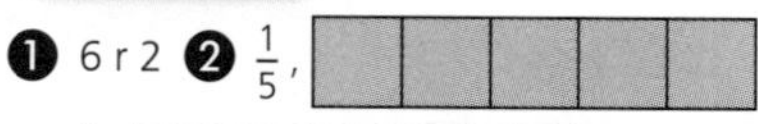

❸ $\frac{3}{4}$, 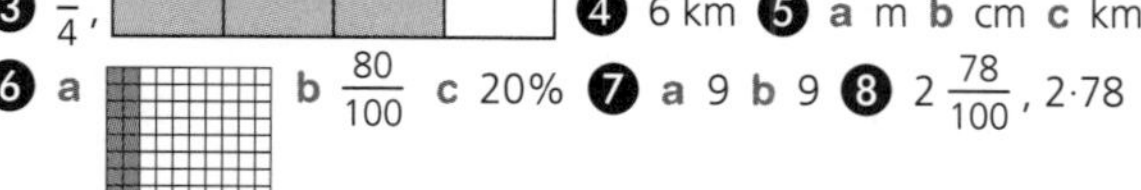 ❹ 6 km ❺ a m b cm c km d mm

❻ a b $\frac{80}{100}$ c 20% ❼ a 9 b 9 ❽ $2\frac{78}{100}$, 2·78

❾ 32 ❿ 23 ⓫ 400 ⓬ 6 past 7

6:4

❶ a 1 136 000 b 1 101 464 ❷ 72 ❸ \$70 000

❹ a 10 and 3 b 12 and 2 ❺ 16 ❻ 12 ❼ a 117 b 53

Challenge

Answers may vary.

E.g. It has 7 digits. It is an even number. It has 6 millions, 534 thousands, 1 hundred and 8 ones. It becomes 7 million when rounded to the nearest million.

Activity

90, 27, 18, 45, 81, 63, 99, 36, 54, 72

2, 4, 6, 3, 8, 5, 7, 9, 1, 10

7:1

❶ 7 ❷ 7 ❸ 5 ❹ 6 ❺ 87 ❻ 30 ❼ 30 ❽ 32

❾ a $\frac{6}{10}$ b $\frac{2}{5}$ ❿ 8000 m ⓫ $2\frac{1}{4}$

⓬

0·75	**$\frac{75}{100}$**	**75%**

0·20	**$\frac{20}{100}$**	**20%**

⓭ $\frac{7}{4}$ ⓮ a m b mm c cm d km ⓯ 24 cm ⓰ 0·2 ⓱ 10

⓲ 0·43 = 43 out of 100 = 43% ⓳ 5 000 000 000

7:2

❶ 1 ❷ 7 ❸ 7 ❹ 9 ❺ 6 ❻ 8 ❼ 5 ❽ 4

❾ $3\frac{2}{3}$ ❿ a 343 b 622 ⓫ a 4 km 638 m b 8 km 305 m ⓬ $\frac{34}{10}$

⓭ a **50** hundredths b $\frac{50}{100}$ c 50% ⓮ 120 km

⓯ a 0·37, 37% b 0·96, 96% ⓰ 15 ⓱ 83 456 652

Activity

a 20 cm b 24 cm

7:3

❶ $3\frac{3}{4}$ ❷ a $\frac{7}{8}$ b $\frac{2}{10}$ ❸ a 8156 m b 7365 m ❹ 300 km

❺ a **46** hundredths b **46** out of **100** c 0·46 d 46%

❻ $\frac{73}{10}$ ❼ 8 cm ❽ a 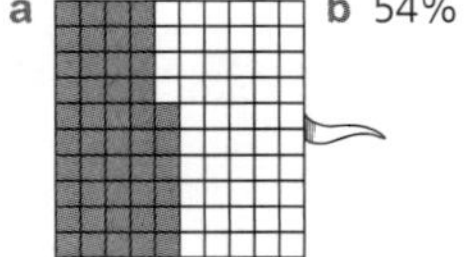 b 54%

❾ a 8000 b 10 000 ❿ a 6000 b 8000

7:4

❶ a \$10 b \$30 ❷ a 61 b 30 ❸ a 10 b 5 c 4 d 2

❹ a $\frac{1}{2}$ or $\frac{5}{10}$ b $\frac{3}{2}$, $\frac{15}{10}$ or $1\frac{1}{2}$

❺ 512, 614 Rule = +102

Challenge

Fraction	Decimal	Percentage
$\frac{20}{100}$	0.20	20%
$\mathbf{\frac{41}{100}}$	0.41	**41%**
$\mathbf{\frac{86}{100}}$	**0.86**	86%
$\frac{38}{100}$	**0.38**	**38%**

Activity

(19) 3 out of 8 (20) $\frac{3}{10}$ shaded (21) 3 tenths shaded (22) 0·3 shaded (23) $\frac{24}{100}$ shaded (24) 0·36 shaded (26) 19 out of 100 (27) 62% shaded (28) 1·36 (30) $0{\cdot}3 = \frac{3}{10}$, $0{\cdot}30 = \frac{3}{10}$

8:1

❶ 106 ❷ 108 ❸ 3 ❹ 7 ❺ 35 ❻ 24 ❼ 80 ❽ 24
❾ $3\frac{4}{5}$ ❿ $\frac{2}{4}$ or $\frac{1}{2}$ ⓫ $\frac{2}{6}$ ⓬ ⓭ 9 each
⓮ a 12 b 2 books c 30 ⓯ a 536 454 b 30 453

8:2

❶ 355 ❷ 762 ❸ 16 ❹ 13 ❺ 49 ❻ 8 ❼ 8 ❽ 5
❾ a $1\frac{3}{4}$ b $1\frac{2}{3}$ ❿ a $\frac{5}{4}$ b $\frac{23}{10}$ ⓫ a 10 cm b 6 cm^2
⓬ a 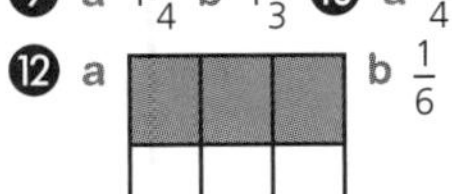 b $\frac{1}{6}$
⓭ 
⓮ $0.65 ⓯ a 58 531 403 b 79·2 c 35·65 ⓰ 4 ⓱ 42, 51, 60, 69

Activity

70, 20, 80, 40, 10, 100, 50, 30, 40, 60
40, 10, 15, 50, 25, 5, 35, 30, 45, 20

8:3

❶ a 10 cm b 4 cm^2 ❷ 35 000 600 ❸ 0·9 ❹ a 70 cm b 124 cm^2 ❺ a 12 h b 61 h ❻ a 56 734 630 b 39 475 279 c 19 273 399
❼ a 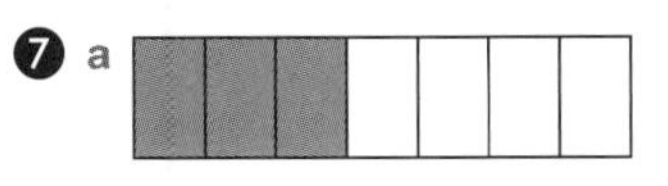 b 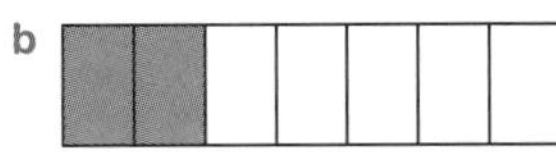
c $\frac{5}{7}$ d $\frac{3}{7}$

8:4

❶ 14, 20 ❷ $10 ❸ a $3\frac{1}{2}$ b $6\frac{1}{2}$ ❹ 50 m^2 ❺ 8 × 6

Challenge

Answers will vary.

Activity

(1) millimetres (2) centimetres (3) metres (4) kilometres (5) millilitres (6) litres (7) grams (8) kilograms (10) minutes (12) square centimetres (13) square metres

9:1

❶ 15 ❷ 21 ❸ 1 ❹ 20 ❺ 1128 ❻ 16 ❼ 28 ❽ 40 ❾ 6
❿ 144 ⓫ a $\frac{4}{6}$ b $\frac{2}{6}$ c $\frac{6}{6}$ ⓬ a A b A ⓭ a $\frac{1}{2}$ or $\frac{6}{12}$ b $\frac{3}{12}$
⓮ 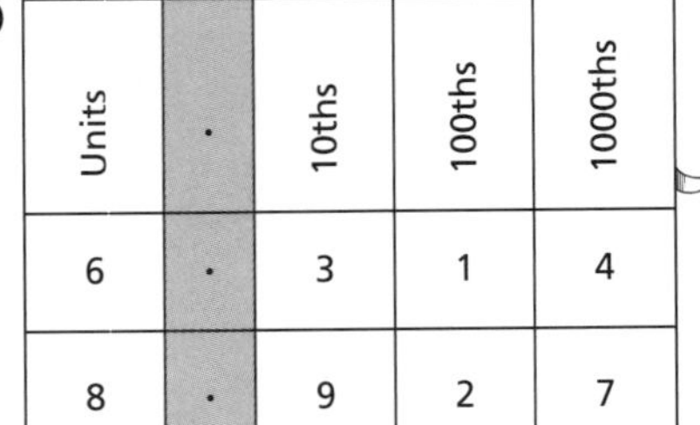

Units	·	10ths	100ths	1000ths
6	·	3	1	4
8	·	9	2	7

9:2

❶ 40 ❷ 35 ❸ 54 ❹ 408 ❺ 826 ❻ 56 ❼ 4 ❽ 4 ❾ 6
❿ 479 ⓫ 1 or $\frac{4}{4}$ ⓬ 15 m^2 ⓭ a 71 m^2 b 24 m^2
⓮ Jessica's toys

Dolls
Games
Tops
0 4 8 12

⓯ a Answers will vary, e.g. 1·115. b Answers will vary, e.g. 1·14. c Answers will vary, e.g. 1·9. ⓰ 34·56

Activity

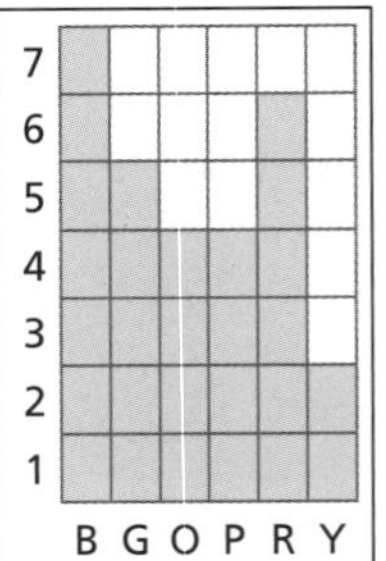

a Blue b 5 c 2

9:3

❶ 1 or $\frac{4}{4}$ ❷ a 32 cm b 64 cm^2
❸ a December b The bikes may be Christmas gifts. c 75
❹

Units	·	10ths	100ths	1000ths
9	·	6	4	2
3	·	8	1	
6	·	0	7	9

❺ 67·471 ❻ m^2

9:4

❶ $\frac{11}{24}$ ❷ 44 m^2 ❸ 87 ❹ before midday (or before noon)
❺ 89 and 99 hundredths will be circled. ❻ 1 h 28 min
❼ a 72 m^2 b $864

Challenge

Answers will vary.

E.g. 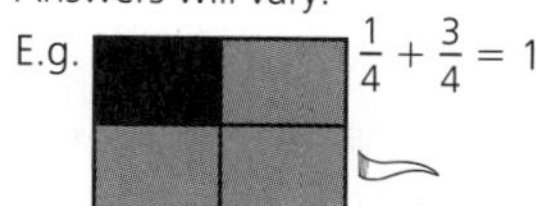 $\frac{1}{4} + \frac{3}{4} = 1$

Activity

54 m

10:1

❶ 49 ❷ 145 ❸ 277 ❹ 63 ❺ 940 ❻ 24 ❼ 412 ❽ 10
❾ 6 ❿ 575 ⓫ **a** $\frac{1}{4}$ **b** $\frac{1}{4}$ ⓬ 1300 m^2

⓭

Chickens	卌 III
Pigs	IIII
Cows	II

⓮ **a** $\frac{3}{4}$ **b** $\frac{1}{4}$ **c** $\frac{1}{4}$ ⓯ 0·46 ⓰ 600

10:2

❶ 40 ❷ 583 ❸ 148 ❹ 54 ❺ 1002 ❻ 34 ❼ 5 ❽ 36
❾ 26 ❿ 407

⓫ 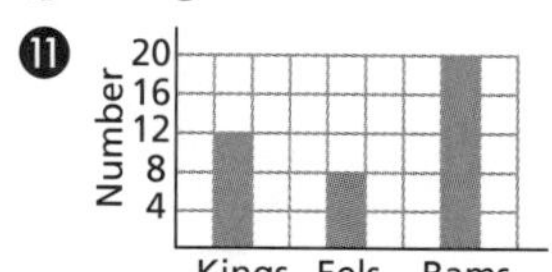

⓬ **a** $\frac{7}{10}$ **b** $\frac{8}{10}$ **c** 1 or $\frac{10}{10}$ **d** $\frac{8}{10}$

⓭ 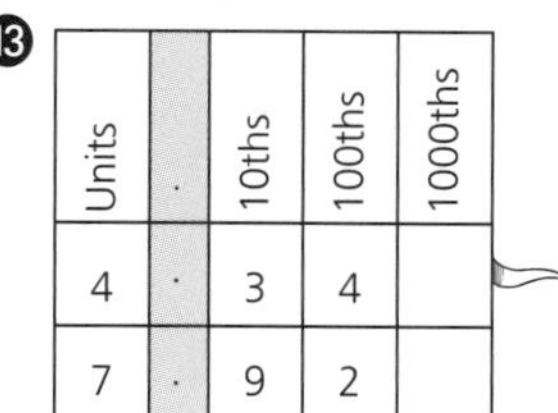

Units	.	10ths	100ths	1000ths
4	.	3	4	
7	.	9	2	

⓮ 2·5 2·6 2·7 2·8 2·9
2·55 2·65 2·75 2·85
⓯ 208

Activity

❶ **a** 1200 **b** 1050 **c** 1053 ❷ **a** 900 **b** 1050 **c** 1047
❸ **a** 6 **b** 300

10:3

❶ **a** $\frac{2}{6}$ **b** $\frac{4}{6}$ **c** $\frac{4}{6}$ ❷ **a** 9 **b** 12 ❸ $\frac{8}{10}$
❹ **a** 36·56 **b** 83·264
❺ 7·3 7·4 7·5 7·6
7·35 7·45 7·55
❻ **a** 56 **b** 3 **c** 6
❼ 4 ½ h ❽ **a** 6:00 pm **b** 9:00 pm ❾ **a** 103 **b** 862

10:4

❶ 14 × 7 ❷ **a** 600 **b** 1440 **c** 201 ❸ **a** 904 **b** 1390
❹ **a** 218 **b** 319 ❺ ◯ ❻ **a** 4539 **b** 3746 ❼ **a** 4238 **b** 2357

Challenge

Answers will vary. e.g.

a 571 + 350, 570 + 351, 467 + 454, etc. **b** 270 + 460, 565 + 165, etc.
c 1000 + 3593, 1093 + 3500, etc.

Activity

a 352 **b** 681

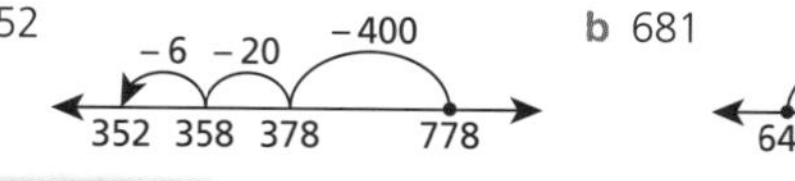

11:1

❶ 388 ❷ 757 ❸ 312 ❹ 211 ❺ 823 ❻ 362 ❼ 647
❽ 475 ❾ 462 ❿ 934 ⓫ **a** 6440 **b** 6000 ⓬ 25 min
⓭ **a** 161 + 300 = 461 **b** 162 + 300 = 462 ⓮ Wednesday
⓯ **a** 45 min **b** 90 min **c** 180 min ⓰ 600 000
⓱ **a** 6 tenths and 8 hundredths **b** 0·68

11:2

❶ 359 ❷ 432 ❸ 493 ❹ 5381 ❺ 761 ❻ 197 ❼ 297
❽ 1350 ❾ 6302 ❿ 932 ⓫ **a** $\frac{3}{8}$ **b** $\frac{5}{8}$ **c** $\frac{5}{8}$ ⓬ **a** 8400 **b** 8000
⓭ 51 min ⓮ 694 ⓯ **a** 9:43 am **b** 8:55 am ⓰ $\frac{1}{8}$, $\frac{1}{4}$, $\frac{1}{2}$ ⓱ 100

Activity

a 8 **b** 15 **c** 60 **d** 120 **e** 65 **f** 160 **g** XXII **h** CV **i** CCXII **j** LXIII
k LXXV **l** CCCLI

11:3

❶ 5000 ❷ **a** 36 months **b** 4 decades **c** 14 weeks

❸ 

❹ **a** 220 + 500 = 720 **b** 540 + 300 = 840 ❺ $671
❻ **a** 8:00 pm **b** 7:00 am

❼ 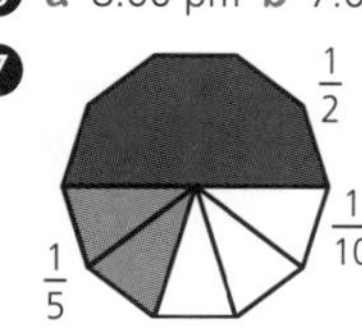

❽ 4 698 557 will be circled. ❾   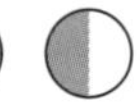

11:4

❶ 2 h 24 min ❷ **a** 6 hours **b** 132 fortnights **c** 12 weeks
❸ **a** 66 **b** 59 ❹ 24 minutes ❺ **a** 120 **b** 1200 ❻ $555
❼ **a** 24 **b** 24

Challenge

Answers will vary.

E.g. **a** 434 – 82, 752 – 400, etc. **b** 415 – 36, 809 – 430, etc.
c 1291 – 487, 3304 – 2500, etc.

Activity

18, 54, 63, 27, 9, 45, 81, 36, 72, 90
9, 18, 27, 36, 45, 54, 63, 72, 81, 90

12:1

❶ 367 ❷ 986 ❸ 522 ❹ 634 ❺ 933 ❻ 76 ❼ 77
❽ 69 ❾ 534 ❿ 332 ⓫ $\frac{1}{10}$, $\frac{1}{5}$, $\frac{1}{3}$ ⓬ a 18:00 b 22:00
⓭ a 161 + 400 = 561 b 361 + 500 = 861 ⓮ 374 cm
⓯ a 1:00 b 12:55 ⓰ 0·39, 0·38, 0·37 ⓱ 4:11 am

12:2

❶ 280 ❷ 249 ❸ 584 ❹ 6491 ❺ 510 ❻ 275 ❼ 459
❽ 2103 ❾ 1040 ❿ 628 ⓫ 411 g ⓬ a $\frac{3}{9}$ b $\frac{6}{9}$ c $\frac{6}{9}$
⓭ a 589 – 400 = 189 b 489 – 200 = 289 ⓮ 1:00 pm
⓯ 9:04 am ⓰ 374 cm ⓱ $\frac{34}{100}$, 0·34

Activity

a 12 cm b 11 cm c 12 cm d 11 cm

12:3

❶ $\frac{9}{10}$ ❷ 1011 g ❸ 6:00 pm ❹ a 18 min b 27 min
c 1 h 10 min or 70 min ❺ a 10:20 pm b 22:20 ❻ 250 m ❼ 7
❽ 8781344, 8780033, 8768367 ❾ 8·1, 8·2, 8·3, 8·4

12:4

❶ 200 mL ❷ $4.20 ❸ 35 ❹ $\frac{1}{10}$
❺ a 14:34 b 20:29 c 02:18 d 05:37
❻ a 20 b 70 ❼ 6 ❽ a 1340 b 1040 ❾ a $3\frac{2}{4}$ or $3\frac{1}{2}$ b $2\frac{4}{8}$ or $2\frac{1}{2}$

Challenge

Answers will vary.

Activity

2, 7, 3, 8, 9, 5, 1, 4, 6, 10
16, 4, 36, 0, 28, 20, 8, 44, 24, 12, 32, 40
56, 28, 49, 7, 77, 0, 35, 70, 14, 21, 42, 63
27, 0, 36, 72, 45, 90, 63, 9, 54, 99, 18, 81

13:1

❶ 900 ❷ 153 ❸ 554 ❹ 700 ❺ 832 ❻ 44 ❼ 8 ❽ 28 mL
❾ 40 ❿ 419 ⓫ 18:00 ⓬ 4, 6, 8, 10, 12, 14
⓭ 3 and 7 will be circled. ⓮ 1, 12, 2, 6, 3, 4
⓯ 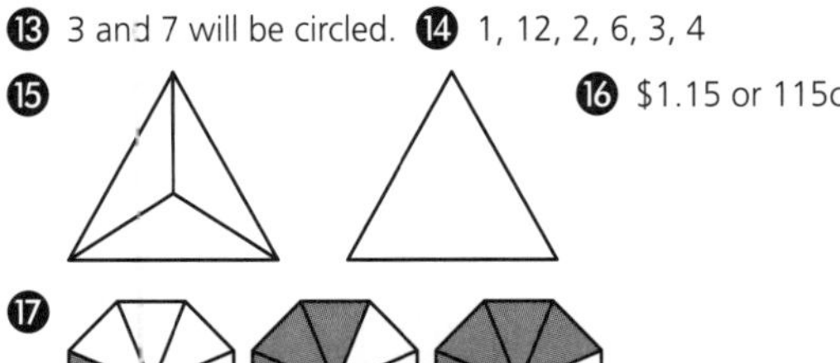 ⓰ $1.15 or 115c
⓱ 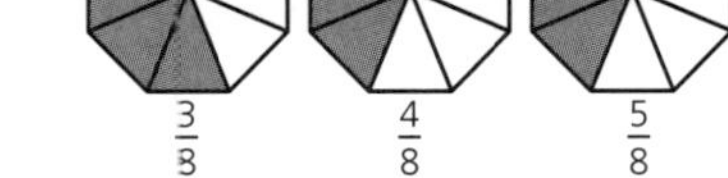

$\frac{3}{8}$ $\frac{4}{8}$ $\frac{5}{8}$

13:2

❶ 14 ❷ 66 ❸ 216 ❹ 759 ❺ 970 ❻ 72 ❼ 8 ❽ 7 m
❾ 7 ❿ 648 ⓫ 4:00 am or 4 am ⓬ a 7:10 pm b 19:10
⓭ a 8, 12, 16, 20, 24, 28 b 16, 24, 32, 40, 48, 56 ⓮ B ⓯ $9.23
⓰ 1, 18, 2, 9, 3, 6 ⓱ $6.55 ⓲ 16 and 8 will be circled.

Activity

a 16 cm b 20 cm

13:3

❶ a 941 – 300 = 641 b 679 – 300 = 379 ❷ 3:47 pm
❸ 12, 18, 24, 30, 36, 42 ❹ 1, 20, 2, 10, 4, 5
❺ octagonal prism ❻ a $5 b $19 c $11 ❼ 6 ❽ $3.13
❾ 19 and 31 will be circled.

13:4

❶ 1, 100, 2, 50, 4, 25, 5, 20, 10 ❷ 60 ❸ $3.15 ❹ $3.30
❺ 7 ❻ $86.76 ❼ 24 ❽ $91 ❾ 28

Challenge

Answers may vary.

a 2, 4, 6, 8, 10, 12, 14, etc.
b 3, 6, 9, 12, 15, 18, 21, etc.
c 4, 8, 12, 16, 20, 24, 28, etc.
d 5, 10, 15, 20, 25, 30, 35, etc.
e 6, 12, 18, 24, 30, 36, 42, etc.

Activity

(4) cube (5) rectangular prism (6) triangular prism (7) hexagonal prism (8) triangular pyramid (9) square pyramid (10) rectangular pyramid (12) cylinder

14:1

❶ $90 ❷ $449 ❸ $34 ❹ 112 ❺ $8.52 ❻ $3.50 ❼ 24 L
❽ 7 ❾ 6 mL ❿ $3.90 ⓫ 6, 9, 12, 15, 18, 21
⓬ This is a triangular prism. It has 5 faces, 9 edges and 6 corners.
⓭ a 650c b $3.75 ⓮ $9.40 ⓯ 1, 28, 2, 14, 4, 7
⓰ no ⓱ The 2nd and 4th shape will be circled.
⓲ 

14:2

❶ 54 ❷ 24 ❸ 56 ❹ 48 ❺ $10.40 ❻ $129 ❼ $848
❽ 210 ❾ 7 g ❿ $2.86 ⓫ 1, 32, 2, 16, 4, 8
⓬ 29 and 19 will be circled. ⓭ 16, 24, 32, 40, 48, 56
⓮ a $2\frac{3}{10}$ b $6\frac{3}{4}$ ⓯ a rectangle or parallelogram
b parallelogram or rhombus ⓰ $1.65
⓱ 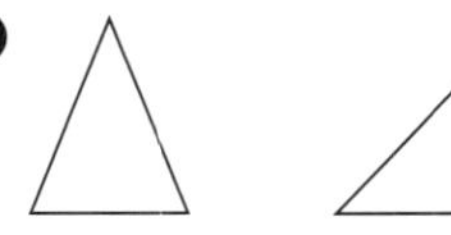 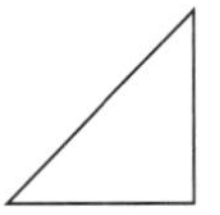

Activity

	÷7	÷35	÷5
35	5	1	7

	÷2	÷4	÷10
40	20	10	4

	÷6	÷24	÷8
48	8	2	6

	÷6	÷2	÷1
36	6	18	36

14:3

1 $5.28 **2** $1.48 **3** a 5 b 8

4 flip — reflection, slide — translation, turn — rotation **5** 1, 40, 2, 20, 4, 10, 5, 8

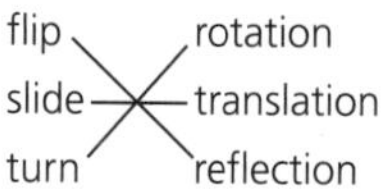

6 a $2\frac{5}{6}$ b $5\frac{1}{4}$ **7** 18, 27, 36, 45, 54, 63

8 a no b yes c yes **9** 24·466 **10** 7 536 096 **11** a 4 b 5

14:4

1 $0.80 **2** a left b right **3** 1, 2, 3, 4, 6 and 12

4 a 14 b 104 **5** 221, 246, 271 **6**

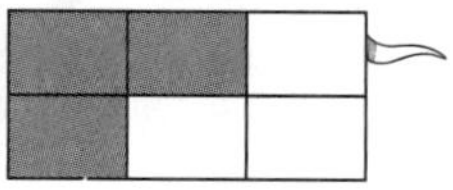

Challenge

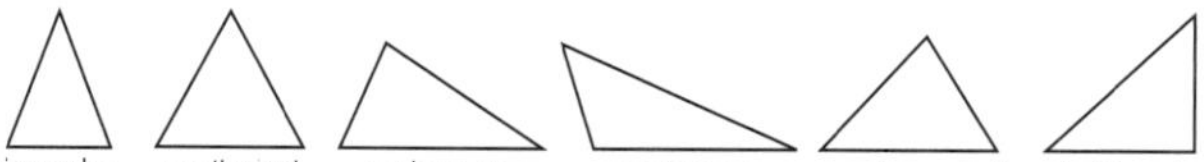

isosceles triangle | equilateral triangle or acute-angled triangle | scalene or acute-angled triangle | scalene or obtuse-angled triangle | acute-angled triangle | right-angled triangle

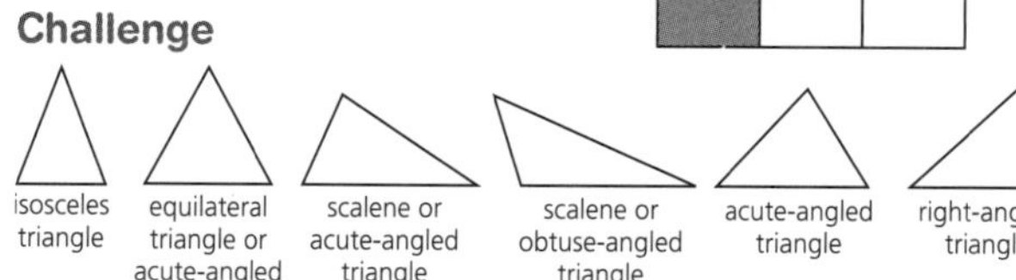

Activity

a $20 b $17.70 c $21 d $33.35 e $18.95 f $46.95 g $52.50

15:1

1 5 **2** 4 **3** 4 **4** 3 **5** $6.85 **6** 5 **7** 5 **8** 4 **9** 10

10 $0.88 **11** 40 000 **12** yes

13 Number of groups = 4, number left over = 2

14 a 14 toys b 7 toys **15** **5** and **6** left over.

16 1400, 1500, 1600

17 a Answers may vary. A square, rectangle, parallelogram or rhombus could be drawn.

b A triangle with one obtuse-sized angle will be drawn.

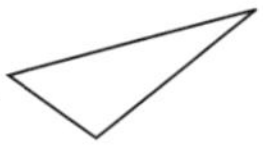

15:2

1 7 **2** 7 **3** 8 **4** 4 **5** $6.79 **6** 9 **7** 6 **8** 6 r 4

9 6 r 5 **10** $4.64 **11** The 1st triangle will be crossed, and the 3rd triangle will be circled. **12** 8 **13** a 7 b 2 **14** a cube

15 a 1000 g b 1000 mL **16** C **17** 8 × 6 + 5 (Answers may vary.)

Activity

(2) parallel lines (3) perpendicular lines (4) horizontal line (5) vertical line (14) tessellation (15) flip (16) slide (17) turn (18) axis of symmetry (19) axes of symmetry

15:3

1 $4.65 **2** $2.15 **3** a 6 r 4 b 9 r 5 **4**

5 One share = 4, left over = 3 **6** a 47 = 7 × 6 + 5 (Answers may vary.) b 33 = 10 × 3 + 3 (Answers may vary.) c 59 = 9 × 6 + 5 (Answers may vary.) **7** a rectangle b 3A c 2B

8 a 3000 m b 400 cm **9** $\frac{1}{10}$, $\frac{1}{5}$, $\frac{1}{3}$

15:4

1 8 **2** Wednesday **3** a $97 b $873 **4** a 51 b 12

5 a $4 b $8 **6** a about 11 or 12 cm b about 4 or 5 cm

Challenge

Answers may vary.

E.g. It is an even number. It has 7 digits. The number after it is 6 534 109. It becomes 7 000 000 when rounded to the nearest million.

Activity

a 4, 8, 12, 16, 20 b 10, 20, 30, 40, 50 c 4, 8, 12, 16, 20

d 10, 20, 30, 40, 50 e 0, 0, 0, 0, 0 f 4, 7, 3, 9, 6

16:1

1 28 **2** 7 **3** 30 **4** 6 **5** 164 **6** 2 **7** 4 **8** 4

9 6 **10** 472 **11**

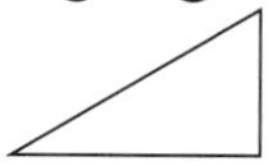

12 false **13** 14 **14** 8

15 a butterfly b cube c umbrella **16** 75°

16:2

1 72 **2** 48 **3** 5 **4** 9 **5** 335 **6** 75 **7** 114 **8** 257

9 9 **10** 428 **11** One share = 6, left over = 3

12 square pyramid **13** a 51 = 5 × 10 + 1 (Answers may vary.)

b 67 = 7 × 9 + 4 (Answers may vary.)

c 73 = 9 × 8 + 1 (Answers may vary.) **14** 68 **15** a 2 b 4

Activity

(1) face (2) vertex or corner (3) edge (11) base (15) net of a cube (16) net of a square pyramid (17) net of a cylinder (18) net of a cone (19) net of a triangular prism

16:3

1 157 **2** 137 **3** 377 **4** 5 r 4 **5** 9 r 4

6 0·11, 0·12, 0·13 **7** One share = **3**, left over = **3**

8 a

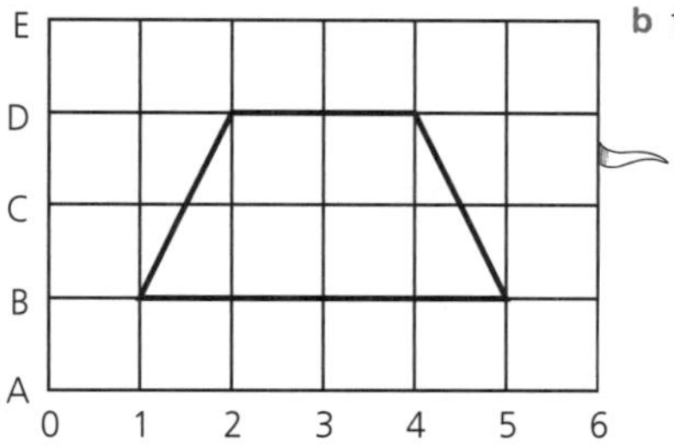

b trapezium

9 a 60 squares will be coloured red and 30 squares will be coloured blue.

b $\frac{1}{10}$ c 0·1 **10** 70°

16:4

1 $169 **2** 90 **3** a 6 b 27 **4** 61 **5** 280

Challenge

Answers will include: an acute angle, right angle, obtuse angle, straight angle, reflex angle, and a revolution.

Activity

a 6 coins each, with 2 left over.

b 6 cards each, with 3 left over.

17:1

1 369 **2** 283 **3** 32 **4** 24 **5** 156 **6** 3 **7** 7 **8** 5

9 35 **10** 8834 **11** 130° **12** 4 **13** a 18 b 17

14 a $\frac{4}{8}$ or $\frac{1}{2}$ b $\frac{2}{8}$ or $\frac{1}{4}$ c $\frac{5}{8}$ d $\frac{1}{8}$

15 6 tens, 7 ones, 3 tenths, 5 hundredths and 6 thousandths.

16 a 0·17 will be circled. b 0·09 will be circled. **17** $3\frac{4}{5}$

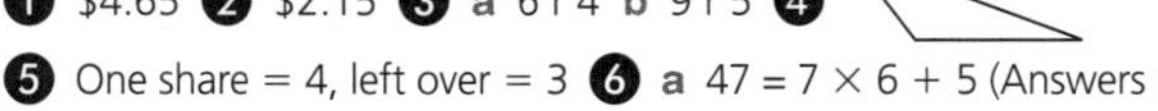

17:2

1 64 2 48 3 54 4 5 5 362 6 652 7 289

8 9 9 9 10 5948 11 \$2.65 12 acute angle

13 8501 mL or 8 L 501 mL 14 **a** 26 000 000 **b** 35 000 000 000

15 **a** 5500 g **b** 25 mm 16 **a** 225 **b** 13

17 8 tens, 2 ones, 2 tenths, 9 hundredths and 3 thousandths.

18 **a** **b**

Activity

12, 6, 21, 15, 27, 18, 3, 24, 9, 30

12, 24, 18, 0, 30, 36, 54, 48, 42, 60

17:3

1 184 2 203 3 107 4 3 r 5 5 8 r 4 6 9201

7 9032 8 \$2.24 9 20° 10 8441 g or 8 kg 443 g

11 **a** 12 **b** no **c** 25 12 0·09 or $\frac{9}{100}$

17:4

1 \$1045 2 **a** 4 **b** 2 3 5, 2 4 30°

5 31, 63, 127, 255 6 32

Challenge

Answers may vary.

E.g. It is an odd number. It is a decimal. It has 8 tens, 3 ones, 5 tenths, 6 hundredths and 1 thousandth.

Activity

1, 9, 5, 4, 0, 8, 2, 6, 3, 7 even – odd = odd

5, 1, 7, 5, 2, 10, 4, 8, 3, 9 odd – odd = even

18:1

1 24 2 5 3 3 4 32 5 7949 6 589 7 762

8 295 9 9 10 1993 11 yes

12 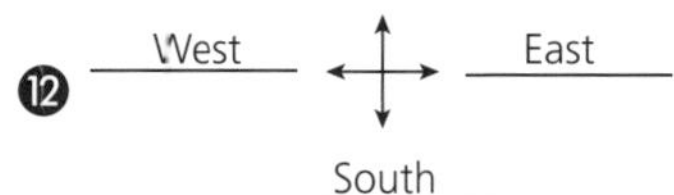

13 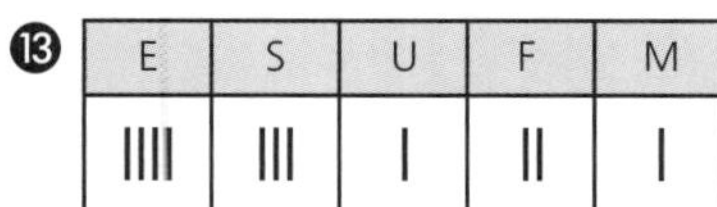

E	S	U	F	M
IIII	III	I	II	I

14 0·14 will be circled. 15 0·67 16 $\frac{52}{100}$ 17 yes 18 0·03 or $\frac{3}{100}$

19 **a** **b**

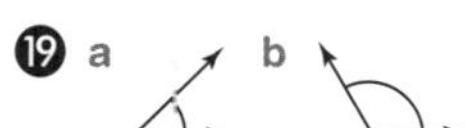

18:2

1 6 2 49 3 962 4 54 5 \$84.01 6 6 7 7 8 443

9 526 10 1807 11 **a** A **b** C 12 **a** acute angle **b** obtuse angle

13 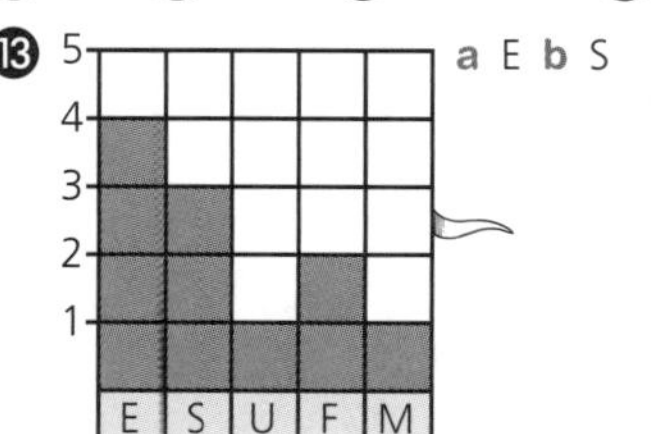

a E **b** S

14 **a** a car **b** a person **c** a bus **d** a house

Activity

a south-east **b** south-west **c** north-west

18:3

1 1563 + 1, 1564 2 3502 + 1, 3503

3 **a** C, B, D, A **b** reflex angle **c** right angle

4 **a** straight angle **b** acute angle **c** obtuse angle **d** acute angle

5 **a** $\frac{7}{12}$ **b** $\frac{11}{12}$ **c** $\frac{6}{12}$ **d** $\frac{6}{12}$ 6 \$24 7 1, 32, 2, 16, 4, 8 8 $\frac{447}{100}$

18:4

1 **a** 4 times **b** 2 times 2 3 3 $22\frac{1}{2}$° or 22·5°

4 **a** 93 **b** 81 5 497 6 500 000

Challenge

a 2463 **b** 3709

Activity

8, 32, 48, 24, 40, 64, 16, 72, 56, 80

6, 24, 36, 18, 30, 48, 12, 54, 42, 60

19:1

1 21 2 48 3 16 4 32 5 \$84.07 6 5 7 4

8 6 9 8 10 2102 11 yes 12 14 13 9 000 000

14 **a** B2 **b** D3 **c** dam **d** stable 15 1, 12, 2, 6, 3, 4

16 200 17 16 18 24, 30, 36, 42, 48

19:2

1 45 2 40 3 72 4 42 5 \$66.45 6 5 7 8

8 7 9 8 10 5056 11 0·14 will be circled. 12 4

13 2 rectangles will be shaded. 14 4 15 1 900 247 16 7 000 000

17 $\frac{5}{8}$ 18 **a** E = 100 + 100 = 200 **b** E = 600 − 300 = 300 19 $\frac{3}{6}$ or $\frac{1}{2}$

Activity

1 **a** dog **b** fish **c** cupcake **d** bird

2 **a** M **b** P **c** P **d** C

19:3

1 3250 2 5750 3 6957 4 41 5 23 6 12

7 662 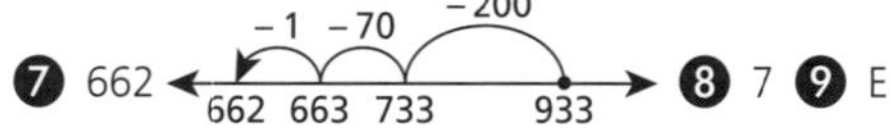

8 7 9 E

10 **a** 45 000 000 **b** 37 000 000 000

11 **a** $\frac{5}{10}$ **b** $\frac{4}{5}$ **c** $\frac{4}{10}$ **d** $\frac{5}{5}$

19:4

1 $\frac{1}{2}$ each 2 6 3 23 4 **a** 154 **b** 28 5 46 6 Monday

7 **a** 75° **b** 285° 8 4:30 pm

Challenge

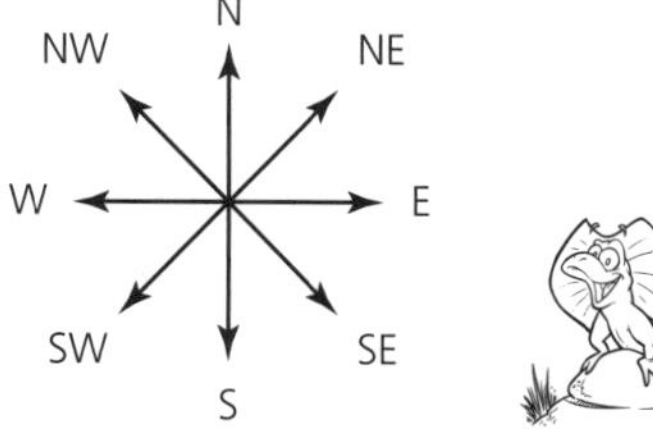

Activity

❶ a C b A ❷ B ❸ 14

20:1

❶ 60 ❷ 32 ❸ 36 ❹ 25 ❺ 7525 ❻ 10 ❼ 5 ❽ 8 ❾ 13
❿ $5.06 ⓫ 2 hundreds, 1 tens, 6 units, 9 tenths, 3 hundredths
⓬ a C b 4 ⓭ red ⓮ 3 333 568 ⓯ 1, 24, 2, 12, 3, 8, 4, 6

20:2

❶ 63 ❷ 64 ❸ 35 ❹ 36 ❺ 8132 ❻ 81 ❼ 9 ❽ 7
❾ 8 ❿ $16.78 ⓫ 103 ⓬ 12 ⓭ 32 ⓮ a yes b C
⓯ a 3 b 3 ⓰ a 4640 b 5000 ⓱ a 5 b 3 c 40 d 3
⓲ a 0·3 b 0·7 c 0·5 d 1·1

Activity

0, 3, 7, 8, 6, 4, 2, 5, 1, 9
7, 5, 9, 2, 8, 6, 0, 10, 3, 1
6 + 9 = 15
8 + 6 = 14

20:3

❶ 4679 ❷ 4630 ❸ 179 ❹ 21 ❺ 22 ❻ 13 ❼ 123
❽ 423 ❾ 151 ❿ a B b A ⓫ yes ⓬ 15
⓭ 4 hundreds, 3 tens, 4 units, 5 tenths, 8 hundredths.
⓮ a 90 b 180 c 360

20:4

❶ 16 ❷ a $\frac{1}{2}$ b $\frac{1}{4}$ ❸ 2 April 2024 ❹ 2 ❺ 9 ❻ 4079 ❼ 85

Challenge

Activity

Is this game fair? No. Answers will vary for the tally.

21:1

❶ 100 ❷ 340 ❸ 6400 ❹ 530 ❺ 10 307 ❻ 70 ❼ 2300
❽ 300 ❾ 2100 ❿ $19.89 ⓫ 60
⓬ a E = 200 + 100 = 300 b E = 800 − 300 = 500
⓭ a 500 g b 250 mL ⓮ yes ⓯ a $\frac{13}{5}$ b $\frac{10}{3}$
⓰ 2·37 ⓱ 37 000 000 ⓲ 8 ⓳ $5\frac{2}{3}$

21:2

❶ 5800 ❷ 2800 ❸ 490 ❹ 420 ❺ 9200 ❻ 140
❼ 180 ❽ $560 ❾ 6300 ❿ $53.94 ⓫ 25 cm ⓬ $9
⓭ a acute b obtuse ⓮ 302 ⓯ a 5·7 L b 8·451 kg c 6·7 cm
d 3·75 m ⓰ 46 000 000 000 ⓱ a $\frac{19}{100}$, $\frac{18}{100}$, $\frac{17}{100}$
b Subtract 1 hundredth (or subtract $\frac{1}{100}$). ⓲ 14 cm

Activity

14, 42, 7, 28, 70, 49, 21, 63, 35, 56
16, 48, 8, 32, 80, 56, 24, 72, 40, 64

21:3

❶ 2543 ❷ 2815 ❸ 1316 ❹ 9 r 1 ❺ 5 r 2 ❻ 4 r 3
❼ 302 ❽ 324 ❾ 130 ❿ E = 400 + 0 = 400
⓫ E = 500 − 200 = 300 ⓬ no ⓭ 24°C ⓮ 240
⓯ a 27 b 3500 c 4876

⓰

0·40	$\frac{40}{100}$	40%

0·95	$\frac{95}{100}$	95%

⓱ 12 ⓲ 5

21:4

❶ 146 cm ❷ 10 and 11 ❸ a 48 b 25 ❹ a 52 r 1 b 31 r 2
❺ 9 o'clock and 3 o'clock. ❻ a 810 b 200 c 3200 d 1000

Challenge

Answers will vary.
E.g.
Rule: Add 0·02
Pattern: 0·35, 0·37, 0·39, 0·41
Rule: Subtract $\frac{5}{100}$.
Pattern: $\frac{73}{100}$, $\frac{68}{100}$, $\frac{63}{100}$, $\frac{58}{100}$

Activity

45, 27, 9, 36, 18, 63, 90, 81, 72, 90
42, 14, 56, 28, 35, 7, 49, 21, 63, 70

22:1

❶ 240 ❷ 80 ❸ 3500 ❹ 320 ❺ 9751 ❻ 8100
❼ 350 ❽ 5 ❾ 4 ❿ $13.64 ⓫ obtuse angle ⓬ 6000
⓭ 5 out of 8 ⓮ $\frac{2}{3}$ ⓯ 10 ⓰ a 47 b 29 ⓱ a $\frac{2}{6}$ b $\frac{12}{15}$
⓲ $\frac{11}{2}$ ⓳ 5, $5\frac{1}{2}$, 6 ⓴ square pyramid

22:2

❶ 400 ❷ 5400 ❸ 4900 ❹ 3600 ❺ 7150 ❻ 2400
❼ 4500 ❽ 9 ❾ 7 ❿ $45.59 ⓫ 1·4 ⓬ 12 s
⓭ $\frac{11}{100}$, $\frac{15}{100}$, $\frac{21}{100}$ ⓮ 3 ⓯ a $\frac{2}{8}$ or $\frac{1}{4}$ b $\frac{5}{8}$ c $\frac{7}{8}$ d $\frac{1}{8}$
⓰ $2\frac{4}{5}$ ⓱ a $\frac{6}{12}$ b $\frac{3}{12}$ ⓲ cube

Activity

The placement of numbers may vary but each line will have a total of 18.

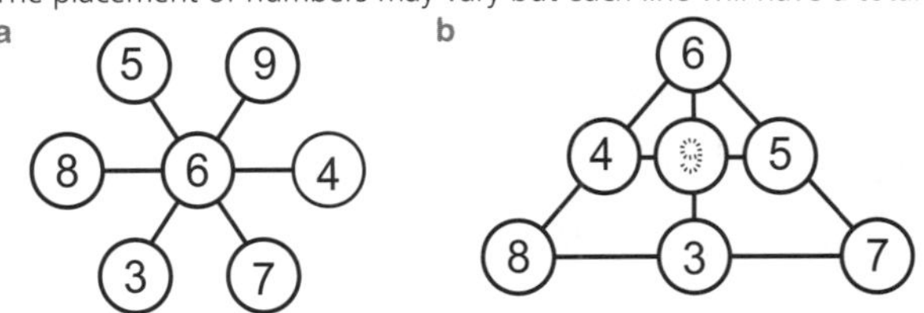

22:3

1 5369 **2** 3778 **3** 6576 **4** 2 r 4 **5** 6 r 1

6 92 **7** 120 **8** 182 r 3 **9** 63

10

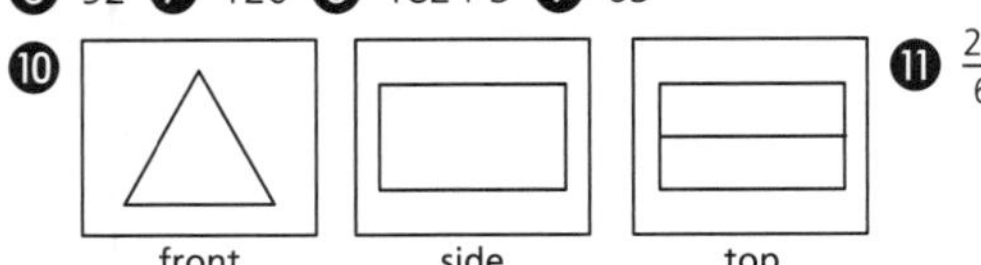

11 $\frac{22}{6}$

12

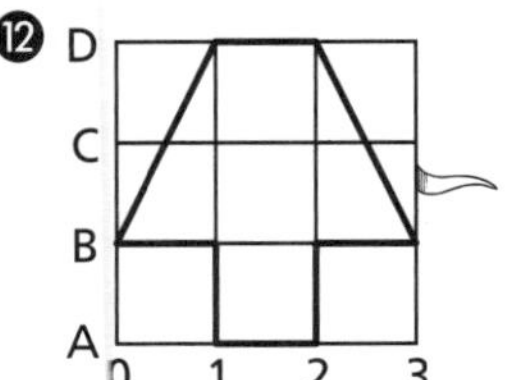

13 a $\frac{4}{8}$ b $\frac{2}{10}$ **14** $2\frac{5}{4}$ **15** a $\frac{6}{12}$ b $\frac{3}{12}$ **16** a 9000 m b 40 mm

22:4

1 a 7 b 10 c A **2** a 0·1 b 0·07 c 0·35 d 0·14

3 a 5·8, 6·9, 8·0, 9·1 b 1·4, 1·6, 1·8, 2·0 **4** 120

Challenge

Answers will vary; a design with rotational symmetry will be shown.

Activity

2, 6, 3, 9, 7, 4, 10, 8, 5, 1
2, 5, 8, 1, 7, 9, 3, 6, 4, 10

23:1

1 116 **2** 169 **3** 125 **4** 12 **5** 9900 **6** 16 **7** 40

8 2 **9** 6 **10** $20.63 **11** $\frac{15}{2}$ **12** a $\frac{3}{15}$ b $\frac{4}{12}$

13 trapezium

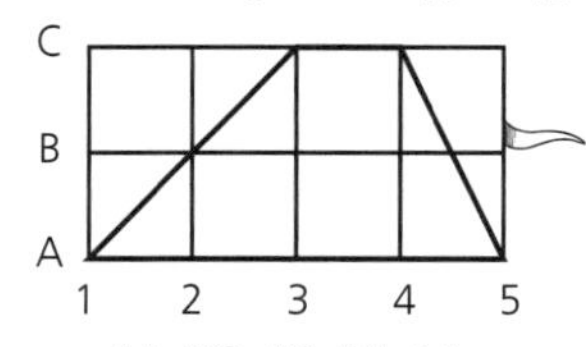

1A, 3C, 4C, 5A, 1A

14 cylinder **15** Estimates will vary. 6·5 cm (or 65 mm) **16** 3 cm 2 mm

23:2

1 72 **2** 30 **3** 35 **4** 54 **5** 6 **6** 4 **7** 506

8 715 **9** 43 r 7 or $43\frac{7}{10}$ **10** 116 r 4 or $116\frac{4}{7}$ **11** 156 r 1 or $156\frac{1}{4}$

12 $3\frac{1}{2}$ **13** cone **14** a $\frac{15}{20}$ b $\frac{10}{20}$ c $\frac{1}{10}$ d $\frac{5}{10}$

15 a Asia b Africa c South America d F2

Activity

(5) rhombus (6) trapezium (7) parallelogram (8) kite (9) quadrilaterals (10) pentagons (11) hexagons (12) octagon (13) regular shapes (14) irregular shapes

23:3

1 7406 **2** 8116 **3** 9520 **4** 8 r 2 or $8\frac{2}{3}$ **5** 7 r 4 or $7\frac{4}{5}$

6 8 r 4 or $8\frac{4}{6}$ **7** $\frac{79}{9}$ **8** no **9** rectangular prism

10 $7\frac{1}{2}$ **11** a $\frac{2}{6}$ b $\frac{8}{10}$

12 a D3 b D2 c C2 d D1 **13** 38°C

23:4

1 a 15 cm b 45 cm **2** 300 mm **3** a 5850 mm b 16 400 mm

4 a 3B b 6B **5** a 250 cm b 25 mm

Challenge

Answers will vary.

Activity

1 B **2** F

24:1

1 179 **2** 45 **3** 28 **4** 24 **5** 27 **6** 450 **7** 10 **8** 10

9 36 **10** 39 r 1 or $39\frac{1}{7}$ **11** 48 r 1 or $48\frac{1}{8}$ **12** 5·03, 5·36, 5·39, 5·7

13 a $\frac{5}{10}$ b $\frac{3}{18}$ **14** yes **15** 4·8 **16** 10 **17** 35 000

18 a m b mm **19** 3·9 cm **20** 656 cm **21** 8:27 am **22** perimeter

24:2

1 0·1 **2** 20 **3** 134 **4** 165 **5** 3500 **6** 2 **7** 774

8 4 **9** 53 r 4 or $53\frac{4}{7}$ **10** 42 r 8 or $42\frac{8}{9}$ **11** 27 r 4 or $27\frac{4}{10}$

12 a $\frac{1}{3}$ b $\frac{4}{5}$ **13** 3·06 m, 3·58 m, 3·6 m, 3·68 m **14** a 3·85 m b 8·36 m

15 yes **16** a 7·4 b 8·30 **17** Estimates will vary. 42 mm

18 a 32 m b 18 cm **19** 2 h 39 min

Activity

a Monday b Tuesday c 8 d

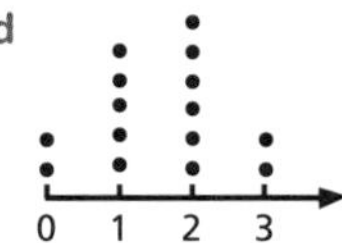

24:3

1 84 **2** 62 r 3 or $62\frac{3}{6}$ **3** 74 r 2 or $74\frac{2}{10}$

4 6·99 L, 26·09 L, 26·59 L, 26·8 L, 31 L

5 no **6** a 3·8 b 3·06

7 a 9·23 9·24 9·25 9·26 / 9·235 9·245 9·255 b 9·245

8 Estimates will vary. 47mm **9** a $\frac{1}{4}$ b $\frac{1}{2}$ c $\frac{3}{8}$ d $\frac{16}{16}$

10 a 5·7 cm b 3·9 cm **11** 36 mm **12** 4:45 pm

24:4

1 8 cm **2** a 10 cm b 30 cm **3** 141·6 cm **4** one quarter

5 Answers will vary. **6** a 12 h 35 min b 15 h 36 min

7 $59.25 **8** a 30, 39, 48, 57 b 62, 81, 100, 119 c 73, 50, 27, 4

Challenge

Answers may vary.

E.g.

a $\frac{2}{4}, \frac{3}{6}, \frac{5}{10}$ b $\frac{2}{6}, \frac{3}{9}, \frac{6}{18}$ c $\frac{2}{8}, \frac{4}{16}, \frac{25}{100}$

Activity

a 18:00 b 09:30 c 07:00 d 22:30 e 12:00 f 24:00 or 00:00

25:1

1 176 **2** 150 **3** 4 **4** 6 **5** 7 **6** 28 **7** 349 **8** 300

9 38 r 5 or $38\frac{5}{10}$ **10** 402 **11** 117 r 2 or $117\frac{2}{6}$ **12** 2·08, 2·46, 2·74, 2·9

13 a $\frac{6}{8}$ b $\frac{8}{12}$ **14** a 7·4 b 6·04 **15** 14 kg **16** 1 m

17 30 min **18** 19:25 **19** 4:47 pm

25:2

❶ 180 ❷ 7200 ❸ 650 ❹ 927 ❺ 100 ❻ 1800
❼ 9 ❽ 745 ❾ 42 r 1 or $42\frac{1}{6}$ ❿ 75 r 4 or $75\frac{4}{5}$ ⓫ 46 r 3 or $46\frac{3}{10}$
⓬ **a** 10 cm **b** 6 cm^2 ⓭ 7·355, 17·09, 17·3, 17·5
⓮ **a** 7·9 **b** 3·06 ⓯ **a** 82 cm **b** 29 cm ⓰ 110 mL ⓱ 02:46

Activity

a 2 **b** 2 **c** 8 **d**

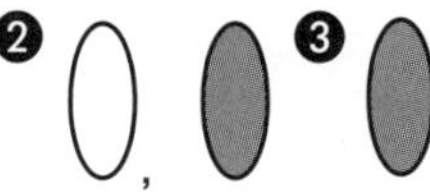

25:3

❶ 85 r 6 or $85\frac{6}{10}$ ❷ 204 ❸ 46 r 3 or $46\frac{3}{10}$ ❹ tonnes
❺ 10·096, 10·323, 10·384, 10·73 ❻ no ❼ **a** 4·1 **b** 8·49
❽ **a** 43 cm **b** 61 cm ❾ 4·546 km ❿ 60 000 ⓫ 15:56
⓬ 7:41 pm ⓭ 16, 8 and 4 will be circled.
⓮ 1, 24, 2, 12, 3, 8, 4, 6 ⓯ 4 kg 656 g ⓰ 8500 ⓱ false

25:4

❶ 9 and 8.
❷ ❸
❹ 89 L ❺ **a** 198 **b** 315
❻ 1, 100, 2, 50, 4, 25, 5, 20, 10 ❼ 184 ❽ 439, 521, 603

Challenge

Answers will vary.
E.g.
g: pencil, eraser, etc.
kg: a human, a chair, etc.
t: a truck, a ship, etc.

Activity

11, 22, 33, 44, 55, 66, 77, 88, 99, 110
33, 77, 55, 22, 88, 11, 66, 99, 44, 110

26:1

❶ 250 ❷ 1800 ❸ 150 ❹ 18 ❺ 70 ❻ 15 ❼ 4
❽ 2 ❾ 100 r 5 or $100\frac{5}{7}$ ❿ 74 r 5 or $74\frac{5}{10}$ ⓫ 101 ⓬ yes
⓭ 7·04, 7·27, 8·09, 8·34 ⓮ 12, 3, 6 and 15 will be circled.
⓯ $4.00 ⓰ 5 kg 367 g ⓱ 1, 28, 2, 14, 4, 7 ⓲ true
⓳ 8 ⓴ **a** 14 cm **b** 6 cm^2 ㉑ 0·35

26:2

❶ 500 ❷ 104 ❸ 56 ❹ 400 ❺ 258 ❻ 60 ❼ 112
❽ 70 ❾ 50 r 3 or $50\frac{3}{10}$ ❿ 408 ⓫ 85 ⓬ 189 ⓭ 222
⓮ 8, 16, 24, 32, 40, 48 ⓯ yes ⓰ 2·005, 2·08, 4·009, 4·02
⓱ 1, 32, 2, 16, 4, 8 ⓲ **a** 500 **b** 1500 ⓳ 55
⓴ true ㉑ **a** 24 cm^2 **b** 28 cm ㉒ 4·07, 4·109, 4·65, 4·804

Activity

4, 2, 8
6, 1, 2, 4
3, 2, 6
12, 4, 6, 3

26:3

❶ $18 \times 4 = 9 \times 2 \times 4 = 72$ ❷ $32 \times 25 = 8 \times 4 \times 25 = 800$
❸ 64 r 8 or $64\frac{8}{10}$ ❹ 14 r 9 or $14\frac{9}{10}$ ❺ 91 r 5 or $91\frac{5}{10}$
❻ 1, 36, 2, 18, 3, 12, 4, 9, 6 ❼ 6, 12, 18, 24 ❽ 4500
❾ 6·007, 6·276, 8·046, 8·105 ❿ 2 kg 967 g
⓫ **B** will be circled. ⓬ 12 centimetres ⓭ true
⓮ 52 ⓯ 3 ⓰ **a** 16 cm **b** 16 cm^2 ⓱ **a** 32 cm **b** 64 cm^2

26:4

❶ 24 m ❷ **a** 276 **b** 135 ❸ 3, 4
❹ 2:12 ❺ **a** 250 cm **b** 25 mm ❻ 16 cm
❼ 930, 1121, 1312, 1503 ❽ $1410\frac{1}{2}$ or 1410·5

Challenge

Answers will vary.

Activity

a 1, 15, 3, 5 **b** 1, 21, 3, 7 **c** 1, 16, 2, 8, 4 **d** 1, 12, 2, 6, 3, 4

27:1

❶ 60 ❷ 160 ❸ 1200 ❹ 180 ❺ 9101 ❻ 1 ❼ 46 ❽ 231
❾ 628 ❿ $30.78 ⓫ 3, 6, 9, 12, 15, 18 ⓬ 15 ⓭ **a** 28 **b** 56
⓮ **a** 8, 12, 16, 20, 24, 28, 32, 36 **b** 7, 12, 17, 22, 27, 32, 37, 42
⓯ 555 ⓰ $4 \times 27 = (4 \times 20) + (4 \times 7) = 108$ ⓱ 5·6 cm
⓲ **a** 5·647 kg **b** 8·342 L ⓳ 47 000 000
⓴ Answers may vary. A triangle with three acute angles and a scalene triangle will be drawn.

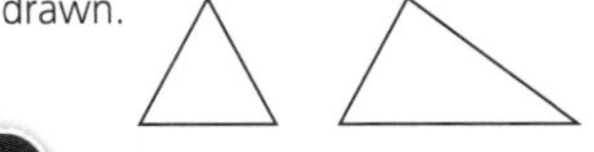

27:2

❶ 480 ❷ 450 ❸ 612 ❹ 264 ❺ 64 ❻ 80 ❼ 862
❽ 311 ❾ 75 ❿ 100 r 4 or $100\frac{4}{7}$ ⓫ 245 ⓬ 68 ⓭ $60
⓮ **a** 50 **b** 100 ⓯ 4, 8, 12, 16, 20, 24
⓰ 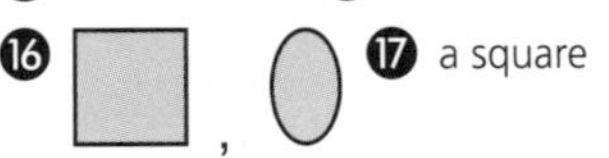 ⓱ a square
⓲ **a** 8000 Estimates may vary. **b** 1993

Activity

2, 1, 5
9, 27, 3
5, 6, 15
21, 7, 6

27:3

❶ $16 \times 5 = 8 \times 2 \times 5 = 80$ ❷ $16 \times 25 = 4 \times 4 \times 25 = 400$
❸ 407 ❹ 403 ❺ 104 ❻ 270 ❼ 9, 18, 27, 36, 45, 54
❽ 41 ❾ 24 ❿ multiply top by 3
⓫ **a** 23 **b** March **c** The netball season may have been finished.

 • *AUSTRALIAN SIGNPOST MATHS NSW 5 MENTALS* • ISBN 978 0 6557 0912 1

27:4

1 a 636 b 6 × 365 = (6 × 300) + (6 × 60) + (6 × 5) = 2190

2 2191 (One of the years will be a leap year.) **3** a 17 b 67

4 a 73 = 7 × 10 + 3 (Answers may vary.)
b 61 = 8 × 7 + 5 (Answers may vary.)
c 89 = 11 × 8 + 1 (Answers may vary.)

5 a 1072 b 2144 **6** a 30 b 30

Challenge

The number 45 is 4 tens and 5 ones. it is an odd number. The number after it is 46 and the number before it is 44. Its factors are 1, 45, 3, 15, 5 and 9.

Activity

a 27, 31, 35, 39, 43
b 4, 8, 16, 32, 64
c 73, 64, 55, 46, 37
d 99, 110, 121, 132, 143
e 7, 15, 31, 63, 127
f 18, 24, 42, 96, 258
37, 32, 27, 22, 17, 12

28:1

1 300 **2** 1 **3** 123 **4** 333 **5** 8882 **6** 6 **7** 4

8 4·92 **9** 8 **10** $8.11

11 a 7 = 3 × 2 + 1 (Answers may vary.)
b 11 = 4 × 2 + 3 (Answers may vary.)
c 19 = 5 × 3 + 4 (Answers may vary.)

12 3 × 46 = (3 × 40) + (3 × 6) = 138 **13** 5, 10, 15, 20, 25, 30

14 $84 **15** a 7 000 000 b 7 464 000

16 3 395 637, 3 392 035, 3 359 574 **17** 3:35 pm

18 $4\frac{2}{4}$ **19** 540 minutes **20** 5 768 000 metres

21 186, 196, 206

28:2

1 70 **2** 340 **3** 655 **4** 720 **5** 7 **6** 9 **7** 7 **8** 472

9 243 **10** 4536 **11** 114 **12** 196 **13** a 1 Oct (or October 1)
b $2\frac{1}{2}$ cm or 2·5 cm **14** 90 **15** 7, 14, 21, 28, 35, 42

16 3 groups, 1 left over **17** 469 days

Activity

Number	Possible factors						
16	(1)	(2)	3	(4)	6	(8)	(16)
25	(1)	2	3	(5)	10	(25)	50
18	(1)	(2)	(3)	4	(6)	(9)	(18)
9	(1)	2	(3)	4	5	6	(9)

Number	Possible factors						
12	(1)	(2)	(3)	(4)	(6)	(12)	24
28	(1)	(2)	(4)	6	(7)	(14)	(28)
81	(1)	(3)	(9)	18	(27)	40	(81)
121	(1)	3	7	9	(11)	22	(121)

28:3

1 728 **2** 171 **3** 430 **4** 331 **5** 431

6 131 **7** 750 **8** 364 **9** 315

10 irregular shapes **11** a 6°C b 10 am

12 6 tens, 7 ones, 3 tenths, 5 hundredths, and 1 thousandth.

13 27·8 cm

28:4

1 3, 12, 21, 30, 39
a + 9 b 75

2

1st number	1	2	3	4	5	6	7
2nd number	45	109	173	237	301	365	429

3 a 4 × 426 = (4 × 400) + (4 × 20) + (4 × 6)
= 1600 + 80 + 24 = 1704
b 9 × 164 = (9 × 100) + (9 × 60) + (9 × 4) = 900 + 540 + 36
= 1476

4 9 o'clock and 3 o'clock will be circled. **5** 7:43 pm

Challenge

2303

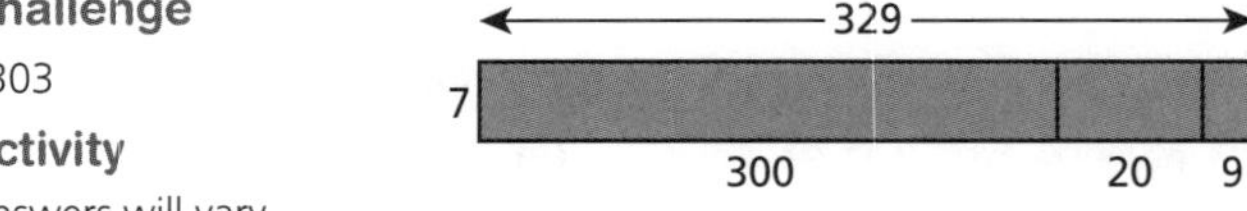

Activity

Answers will vary.

10 kg	2	2	2	0	0	1	1	1
5 kg	1	0	0	4	4	3	2	2
2 kg	0	2	1	2	1	0	2	1
1 kg	0	1	3	1	3	0	1	3

29:1

1 200 **2** 1 **3** 24 **4** 74 **5** 55 **6** 23 **7** 4 **8** 9

9 384 **10** 5173 **11** 4752 **12** a 30 b 41

13 false **14** 5 × 67 = (5 × 60) + (5 × 7) = 335

15 A **16** 1 L or 1000 mL **17** a 6000 b 5700 **18** 18 **19** 4·5 cm

29:2

1 5 **2** 45 **3** 7 **4** 28 **5** 2400 **6** 40 **7** 64 **8** 441

9 185 **10** 2547 **11** $61.28

12 a 8 r 1 b 10 r 1 **13** 2.31 **14** 8 × 32 = (8 × 30) + (8 × 2) = 256

15 a 250 g b 200 g **16** a 4 568 000 b 4 567 900

17 one quarter $\left(\frac{1}{4}\right)$ **18** $7\frac{1}{4}$ or 7·25 **19** 43 **20** 1625 mL or 1 L 625 mL

21 108 months

Activity

(20) 5 × 7 (21) 4 × 10 (22) 20 ÷ 5 (23) 42 ÷ 6 (24) even numbers (25) odd numbers (26) digits (27) 1, 8, 2 and 4 (28) 7 (29) 5

29:3

1 $6.36 **2** 20 547 **3** 13 408 **4** 20 **5** 34 **6** 70 r 3 or $70\frac{3}{10}$

7 2 hundreds, 7 tens, 6 ones, 9 tenths, and 3 hundredths

8 7 × 84 = (7 × 80) + (7 × 4) = 588

9 a 10 b 50 c 11 am to 12 noon d 2 to 3 pm **10** 3

11 a 9 000 000 b 8 976 570

29:4

1 5 days **2** 168 **3** a $\frac{1}{3}$ or $\frac{2}{6}$ b $\frac{3}{8}$ **4** XII **5** 27 **6** 12

Challenge

Answers may vary.

E.g. It is an even number. It has 6 digits. The number after it is 603 731. It becomes 600 000 when rounded to the nearest hundred thousand.

 • *AUSTRALIAN SIGNPOST MATHS NSW 5 MENTALS* • ISBN 978 0 6557 0912 1

Activity

1 a 39 b 40 c 304 d 400 e 621 f 1551

2 a LXV b CCXL c CMLXVI d MMD e CCCXCII f MCMXCIX

30:1

1 40 **2** 597 **3** 206 **4** 13 **5** 7363 **6** 10 **7** 80

8 11 **9** 3 **10** 5782 **11** E = 30 + 60 = 90

12 E = 160 + 40 = 200 **13** 1250 g or 1 kg 250 g

14 a 140 b 98 **15** Answers will vary. $\frac{1}{8}$ has a larger denominator than $\frac{1}{4}$ so the whole has been cut into smaller parts.

16 600 **17** 9 **18** $50 **19** $\frac{2}{8}$ or $\frac{1}{4}$ **20** 3·6 cm

30:2

1 17 **2** 18 **3** 4 **4** 350 **5** 4675 **6** 9 **7** 9

8 9 **9** 451 **10** 14 316 **11** E = 7400 + 100 = 7500

12 E = 3300 − 800 = 2500 **13** 4 **14** VIII

15 1 L 700 mL or 1700 mL **16** 12 **17** 0 **18** 900

19 a < b < **20** 6354 g

Activity

a average length about 19 mm, average height about 5 mm

b average mass 12 kg, average width about 13 mm

c average capacity 950 mL, average height about 14 mm

30:3

1 498 **2** 2597 **3** 504 **4** E = 380 − 50 = 330

5 E = 850 + 360 = 1210 **6** $15.20 **7** IV

8 1382 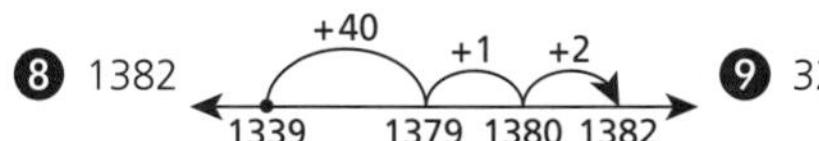

9 32

10 30 × 6 **11** About $140 **12** $\frac{4}{8}$ or $\frac{1}{2}$

13 70 cm **14** a 200 g b 47 mL

30:4

1 $23.55 **2** 73 g, 73 g, 73 g, 99 g

3 a $\frac{5}{12}$ b $\frac{6}{12}$ or $\frac{1}{2}$ c $\frac{1}{12}$ **4** 384 **5** 225

Challenge

Answers will vary.

E.g.
Rule: × 7 − 11
Pattern: 5, 24, 157, 1088, etc.

Activity

18, 42, 12, 36, 54, 6, 30, 0, 24, 48

49, 7, 56, 42, 21, 63, 28, 70, 35, 77

88, 40, 24, 48, 72, 64, 16, 56, 0, 32

31:1

1 55 **2** 11 **3** 24 **4** 34 **5** 6501 **6** 20 **7** 4

8 $34 **9** 79 m **10** $27.86 **11** a 120 minutes b 36 months

12 11 **13** 96 **14** 16, 20, 24, 28 **15** 6000 **16** a $\frac{1}{8}$ b $\frac{7}{8}$

17 1:43 pm **18** 80 **19** $20 will be circled.

31:2

1 4 **2** 33 **3** 8 **4** 5 **5** 40 222 **6** 16 **7** 6 **8** 3000

9 8 **10** $158.52 **11** a $\frac{3}{8}$ b $\frac{5}{8}$ **12** 0·3 **13** $\frac{78}{100}$

14 7·564 L **15** a $\frac{16}{35}$ b yes **16** a 12 b 9

17 $\frac{1}{10}$ **18** 36, 30, 24, 18

Activity

18, 3, 12, 0, 24, 15, 27, 6, 21, 30

30, 45, 15, 35, 10, 50, 25, 40, 20, 0

9, 81, 72, 36, 54, 18, 45, 0, 63, 27

31:3

1 1608 **2** 52 052 **3** 5 **4** $\frac{1}{6}$ **5** 0·33 **6** 5

7 a $\frac{5}{9}$ b $\frac{4}{9}$ **8** a 35 b $\frac{17}{35}$ (head) $\frac{18}{35}$ (tail) **9** 78 000 000

31:4

1 7·8 cm and 8·8 cm. **2** 390 **3** 3 **4** 13

5 a $\frac{1}{2}$ b $\frac{1}{4}$

Challenge

Answers will vary. E.g. flipping a tail on a coin, rolling an even number on a 6-sided dice, spinning yellow on a spinner that is half yellow and half green.

Activity

a false or 2403/

b true or *8(/

c true or *8(/

32:1

1 40 **2** 5 **3** 150 **4** 155 **5** 20 **6** 0·6 **7** 4

8 10 **9** 81 **10** $196 **11** 241 **12** a $\frac{3}{4}$ b $\frac{1}{4}$

13 false **14** litres **15** $\frac{8}{9}, \frac{3}{4}, \frac{1}{3}, \frac{1}{8}$ **16** a $\frac{5}{6}$ b $\frac{3}{4}$

17 6 **18** 9 **19** 60

32:2

1 63 **2** 7 **3** 141 **4** 620 **5** 1·7 **6** 6 **7** 7 **8** 600

9 341 r 1 or 341$\frac{1}{2}$ **10** $99 **11** 100 r 5 **12** a $\frac{3}{10}$ b $\frac{2}{5}$

13 true **14** 06:59 **15** $\frac{1}{5}, \frac{1}{2}, 1\frac{3}{10}, 1\frac{1}{2}$ **16** 4 **17** 9 L 375 mL

18 $\frac{2}{8} + \frac{5}{8} = \frac{7}{8}$ **19** a L b mm c m d m^2

Activity

(6) corner (7) angle (8) acute angle (9) right angle (10) obtuse angle (11) straight angle (12) reflex angle (13) rotation (18) axis of symmetry (19) axes of symmetry

32:3

1 119 r 2 (or 119$\frac{2}{7}$) **2** 153 **3** 77 **4** 14 802 **5** 67 488

6 a $\frac{1}{3}$ b $\frac{4}{10}$ **7** $\frac{1}{4}, \frac{1}{2}, \frac{3}{4}, \frac{7}{4}$ **8** false

9 $\frac{8}{10} - \frac{1}{10} = \frac{7}{10}$ **10** 500 mL **11** 5900 mL

12 7 L 359 mL **13** Milk carton (Answers may vary).

14 a 0·4 will be circled. b 0·71 will be circled.

15 a 52% b $\frac{52}{100}$

32:4

1 a 2500 mL b 0·8 L **2** a 36 b 144 cubes **3** 0·22, 0·24, 0·26

4 a XXVI b LII **5** a 117 b 391 **6** 40

7 Answers may vary.

a 367 + 396 = 363 + 400 = 763

b 847 + 448 = 845 + 450 = 1295

c 925 − 308 = 917 − 300 = 617

Challenge

Answers may vary.

E.g. It is an odd number. It has 7 digits. The number after it is 3 564 010. It becomes 4 000 000 when rounded to the nearest million.

Activity

a 4 b $\frac{3}{4}$ c $\frac{5}{8}$ d $\frac{4}{8}$ or $\frac{1}{2}$ e $\frac{3}{8}$ f $\frac{9}{8}$ or $1\frac{1}{8}$ g $\frac{1}{8}$ h $1\frac{5}{8}$ i $1\frac{3}{4}$ j $\frac{7}{8}$

33:1

1 500 **2** 43 **3** 65 **4** 32 **5** 9221 **6** 6 **7** 3 **8** 3

9 80 **10** 9276 **11** a 1 or $\frac{4}{4}$ b $\frac{3}{4}$ **12** 5000 **13** $\frac{1}{4}, \frac{3}{8}, \frac{1}{2}, \frac{5}{8}$

14 80 **15** 26 500 000 **16** 4 **17** mL **18** $\frac{1}{4}$

19 a 8·3 m b 25% c 1·845 kg d 56% **20** $\frac{6}{8} + \frac{2}{8} = \frac{8}{8}$

33:2

1 305 **2** 498 **3** 122 **4** 8 **5** 4819 **6** 72 **7** 7

8 9 **9** 101 **10** 10 184 **11** a 7500 mm b 4·836 kg

12 64 700 000 **13** a $\frac{1}{5}$ b $\frac{2}{5}$ **14** $\frac{10}{12} - \frac{5}{12} = \frac{5}{12}$ **15** 5500 mL

16 870 mL **17** a 40 cm b $\frac{40}{100}$ or $\frac{4}{10}$ c 40% **18** $44.60

Activity

a

8	3	4
1	5	9
6	7	2

b

9	4	5
2	6	10
7	8	3

c

8	3	10
9	7	5
4	11	6

33:3

1 109 r 2 or $109\frac{2}{5}$ **2** 408 **3** 104 **4** 6164 **5** 34 936

6 a $4\frac{7}{10}$ b $7\frac{6}{10}$ **7** true **8** a 5354 mL b 5·354 L

9 182 days **10** $\frac{1}{8}, \frac{1}{4}, 1\frac{1}{10}, 1\frac{4}{5}$ **11** 94 mL

12 a 23 670 mL b 87% c 3·256 km d 13%

13 15 weeks **14** $54.50 **15** 9·2, 9·3, 9·4, 9·5

33:4

1 a $\frac{4}{5}$ b 80 cm **2** a XXXII b LX **3** a 73 200 000 000

b 69 100 000 000 **4** 6·936 kg or 6 kg 936 g or 6936 g

5 a 13 L b 2730c or $27.30

Challenge

<table>
<tr><td colspan="4">1/2</td><td colspan="4">1/2</td></tr>
<tr><td colspan="2">1/4</td><td colspan="2">1/4</td><td colspan="2">1/4</td><td colspan="2">1/4</td></tr>
<tr><td>1/8</td><td>1/8</td><td>1/8</td><td>1/8</td><td>1/8</td><td>1/8</td><td>1/8</td><td>1/8</td></tr>
</table>

Activity

a $1\frac{1}{4}$ or $\frac{5}{4}$ b $\frac{3}{4}$ c $\frac{1}{4}$ d $\frac{3}{4}$ e $\frac{2}{4}$ or $\frac{1}{2}$ f $\frac{1}{4}$ g $\frac{4}{4}$ or 1

h $\frac{1}{4}$ i $1\frac{1}{2}$ or $\frac{3}{2}$ j $1\frac{1}{4}$ or $\frac{5}{4}$ k $1\frac{1}{2}$ or $\frac{6}{4}$ l $1\frac{1}{4}$ or $\frac{5}{4}$

34:1

1 405 **2** 310 **3** 42 **4** 40 **5** 9827 **6** 3 **7** 5 **8** 4

9 3 **10** 25 480 **11** $3·50 or 350c **12** 4

13 4 × 136 = (4 × 100) + (4 × 30) + (4 × 6) = 400 + 120 + 24 = 544

14 a 0·71, 0·72, 0·73 b Add 0·01. **15** 5 600 000

16 a 35·6% b 7100 mL **17** 5 square kilometres

18 total = $23.75, amount left = $16.25

34:2

1 56 **2** 54 **3** 6 **4** 7 **5** 1008 **6** 8 **7** 9 **8** 63

9 40 **10** 25 488 **11** $\frac{6}{8}, \frac{7}{8}, \frac{8}{8}, \frac{9}{8}, \frac{10}{8}$ **12** $\frac{22}{100}$ **13** 2 900 000 000

14 7·4 L **15** 600 **16** Total = $20.60, amount left = $5.40

17 9 square kilometres

18 5 × 427 = (5 × 400) + (5 × 20) + (5 × 7) = 2000 + 100 + 35 = 2135

Activity

27 friends

34:3

1 a 600 ha b 5 ha

2 58 × 46 = (50 × 40) + (50 × 6) + (40 × 8) + (6 × 8) = 2000 + 300 + 320 + 48 = 2668

3 $36.90 **4** 70%, 50%, 30%, Rule = Subtract 20%

5 6 **6** $\frac{5}{6}$ **7** 7 200 000 **8** a 8·675 km b 3·4 L

9 Answers may vary. a 1800 b 2500 c 4000 d 1200

34:4

1 a 70 000 m² b 2500 ha c 350 000 m²

2 a VIII b LXV **3** a $7\frac{1}{5}, 8\frac{2}{5}, 9\frac{3}{5}, 10\frac{4}{5}$ b Add $1\frac{1}{5}$.

4 $\frac{1}{16}$ **5** a 4964 b 4296

6 Answers may vary. E.g. 150 m × 400 m, 300 m × 200 m.

Challenge

Answers will vary.

E.g.
Rule: × 7 − 11
Pattern: 5, 24, 157, 1088, etc.

Activity

Answers will vary.

35:1

1 673 **2** 277 **3** 42 **4** 35 **5** 48 **6** 49 **7** 6 **8** 4

9 72 × 30 = (70 × 30) + (2 × 30) = 2100 + 60 = 2160

10 400 **11** a 30 000 m² b 7 ha **12** a 6 000 000 m² b 9 km²

13 ha **14** a 4 b green c red d 16 e 30

 • AUSTRALIAN SIGNPOST MATHS NSW 5 MENTALS • ISBN 978 0 6557 0912 1

35:2

1 348 **2** 654 **3** 9 **4** 8 **5** 64 **6** 6 **7** 182 **8** 40

9 $24 \times 20 = (20 \times 20) + (4 \times 20) = 400 + 80 = 480$

10 $58 \times 46 = (50 \times 40) + (50 \times 6) + (40 \times 8) + (8 \times 6)$
$= 2000 + 300 + 320 + 48 = 2668$

11 **a** $16.70 **b** $1.25 **12** 50 000 m^2 **13** 5 km^2

14 **a** 4500 **b** 13 600 or 12 000

Activity

Answers will vary.

35:3

1 $13 \times 64 = 640 + 192 = 832$

2 $5 \times 289 = (5 \times 200) + (5 \times 80) + (5 \times 9) = 1000 + 400 + 45$
$= 1445$

3 7500 m **4** **a** 2400 **b** 18 000 **5** 30 000 m^2

6 **a** 4 000 000 m^2 **b** 8 km^2 **7** $43 \times 52 = (40 \times 52) + (3 \times 52)$

8 **a** 14 **b** 2 **c** 24 **9** 1050 g or 1kg 50 g **10** $1\frac{7}{10}$

35:4

1 **a** 11 **b** 30 matchsticks (when the sides of the hexagons are joined), 36 matchsticks (when each hexagon stands on its own).

2 **a** 3500 **b** 3196 **c** 304

3 Estimates will vary. **a** 900 000 – 1 100 000 km^2 is a good estimate. (The actual area is 984 321 km^2.) **b** 1 500 000 – 1 900 000 km^2 is a good estimate. (The actual area is 1 729 742 km^2.) **4** 23 (Answers may vary)

5 **a** 2356 **b** 1925

Challenge

<table>
<tr><td colspan="3">$\frac{1}{2}$</td><td colspan="3">$\frac{1}{2}$</td></tr>
<tr><td colspan="2">$\frac{1}{3}$</td><td colspan="2">$\frac{1}{3}$</td><td colspan="2">$\frac{1}{3}$</td></tr>
<tr><td>$\frac{1}{6}$</td><td>$\frac{1}{6}$</td><td>$\frac{1}{6}$</td><td>$\frac{1}{6}$</td><td>$\frac{1}{6}$</td><td>$\frac{1}{6}$</td></tr>
</table>

Activity

a 4 **b** 30 s **c** 9 **d** 6 min

36:1

1 1 **2** 280 **3** 180 **4** 150 **5** 9110 **6** 10 **7** 9 **8** 9

9 7 **10** 10 133 **11** $6.50 **12** 60 000 m^2

13 **a** 10 **b** Rachel **c** Heather

14 five million, three hundred and sixty-five thousand, eight hundred and nine.

15 between 180° and 360°

36:2

1 752 **2** 427 **3** 9 **4** 6 **5** 5·49 **6** 70 **7** 6 **8** 8

9 **a** $27 \times 91 = 1820 + 637 = 2457$ **b** $39 \times 17 = 510 + 153 = 663$

10 $3.50 **11** 4 **12** reflex angle **13** 325°

14 **a** 15 m **b** 85% **15** $952

Activity

10, 13, 7, 11, 13, 5, 3, 8, 6, 5

10, 9, 13, 12, 7, 5, 14, 8, 11, 6

36:3

1 253 **2** 903 **3** 2795 **4** Answers may vary. $E = 30 \times 10 = 300$

5 Answers may vary. $E = 40 \times 300 = 12\,000$ **6** 768

7 **a** 120° **b** 60° **8** 4 **9** 4 m 87 cm or 487 cm

10 3270 g **11** **a** $\frac{2}{8}$ **b** $\frac{4}{12}$ **12** nine million, two hundred and four thousand, six hundred and sixty-five. **13** **a** 3 **b** 4

36:4

1 1995 g or 1 kg 995 g **2** **a** 5 **b** 11

3 2 dolls and 2 yo-yos. **4** 14 **5** XCV **6** 98

Challenge

Answers will include:

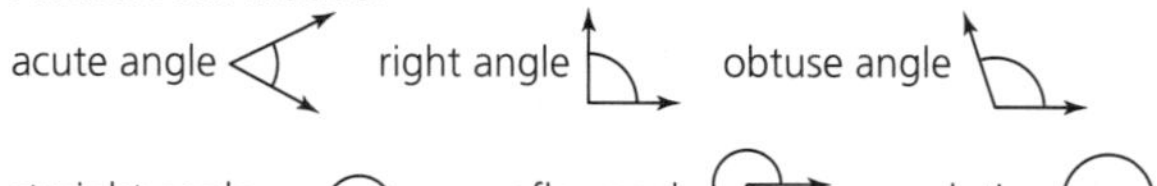

Activity

20, 32, 24, 0, 16, 36, 8, 28, 12, 40

45, 72, 54, 0, 36, 81, 18, 63, 27, 90

37:1

1 240 **2** 2500 **3** 114 **4** 24 **5** 4 **6** 6 **7** 4 **8** 4

9 $30.85 **10** 26 502 **11** 18 102 **12** **a** C **b** D **c** A **d** B

13 **a**

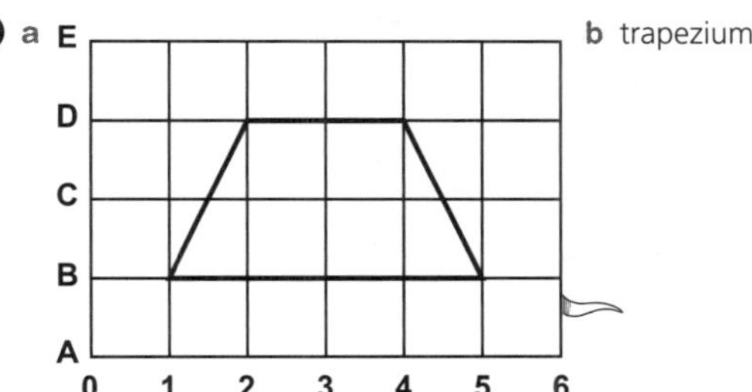

b trapezium

14 nine hundred and three thousand, four hundred and thirteen

15 acute **16** 16:56

37:2

1 3 **2** 4 **3** 24 **4** 56 **5** 6 **6** 4 **7** 7 **8** 10 **9** 2738

10 3818 **11** 1624 **12** $\frac{2}{7}$ **13** 90 000 m^2

14 **a** 5 000 000 m^2 **b** 6 km^2 **15** $684

16 **a** 35° (Estimates between 20–40°)
b 110° (Estimates between 100–120°)
c 200° (Estimates between 190–210°)

17 $\frac{63}{100}$, 60 squares will be coloured red, and 3 squares will be blue.

Activity

a $4^{th} = 10$, $10^{th} = 22$ **b** $4^{th} = 17$, $10^{th} = 41$

37:3

1 462 **2** 780 **3** 732 **4** Answers may vary. $E = 40 \times 10 = 400$

5 Answers may vary. $E = 500 \times 30 = 15\,000$ **6** 3

7 70 000 m^2 **8** 3 km^2

9 **a** a divided bar graph **b** 30 minutes **c** 1 hour **d** 45 minutes
e $1\frac{1}{4}$ hours (or 75 minutes) **10** **a** 70° **b** acute

37:4

❶ a

	3	4	[5]
+	[5]	[1]	1
	8	5	6

b

	[5]	1	8
−	3	[4]	[5]
	1	7	3

❷ $494 ❸ **a** 293 km **b** 330 km ❹ **a** 224 **b** 4

❺ CLX ❻ 194 ❼ 20 cm

Challenge

Answers may vary.

E.g. Blue is the most common favourite colour and Pink is the least common favourite colour. Half of the people have Blue as their favourite colour. One quarter have red as their favourite colour.

Activity

Answers will vary.

Measurement benchmarks

❶ **a** 20 cm **b** 50 cm ❷ **a** 2 m **b** 1·5 m

❸ **a** 4 L **b** 250 mL ❹ **a** 15 kg **b** 2 kg

❺ **a** 2 m^2 **b** 6 cm^2 ❻ **a** 40°C **b** 15°C

 • *AUSTRALIAN SIGNPOST MATHS NSW 5 MENTALS* • ISBN 978 0 6557 0912 1

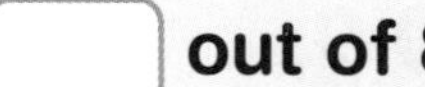

18:3 ☐ out of 8

1
$$\begin{array}{r} 2999+1 \\ \cancel{3000} \\ -1436 \\ \hline +1 \\ \hline = \quad \\ \hline \end{array}$$

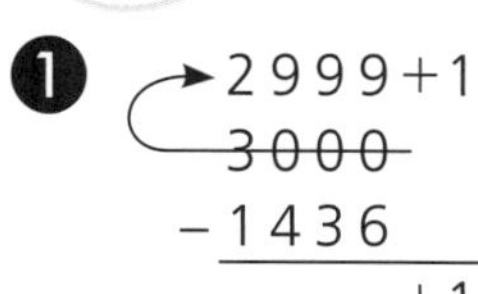

2
$$\begin{array}{r} 5999+1 \\ \cancel{6000} \\ -2497 \\ \hline +1 \\ \hline = \quad \\ \hline \end{array}$$

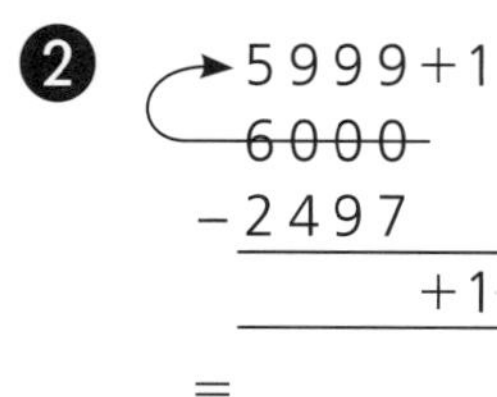

3

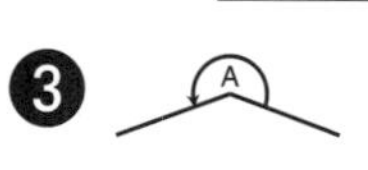

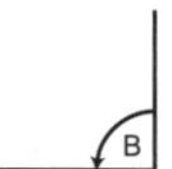

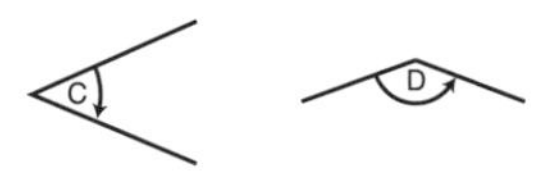

a Put these angles in order, with the smallest first. ________

What type of angle (above) is:

b A? ________ **c** B? ________

4 What kind of angle is shown in each?

a ________ **b** ________

c ________ **d** ________

5

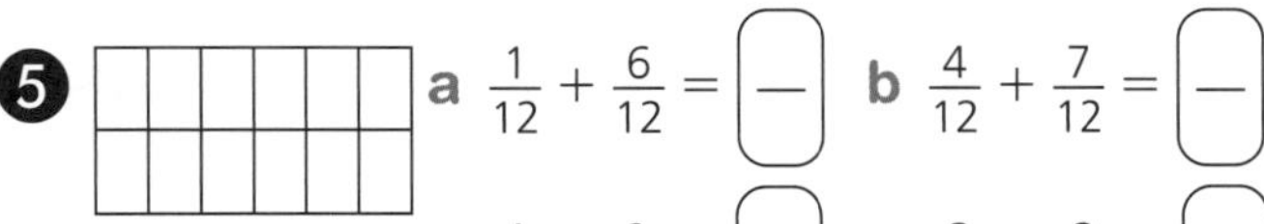

a $\frac{1}{12} + \frac{6}{12} = \frac{\square}{\square}$ **b** $\frac{4}{12} + \frac{7}{12} = \frac{\square}{\square}$

c $\frac{4}{12} + \frac{2}{12} = \frac{\square}{\square}$ **d** $\frac{3}{12} + \frac{3}{12} = \frac{\square}{\square}$

6 Sue wants to save $240 in 10 weeks. How much does she need to save each week? ________

7 Write all the factors of 32. ________

8 Write 4·47 as a fraction. $\frac{\square}{\square}$

18:4 ☐ out of 6

Extension

1 Estimate how many times angle ***X*** would fit into:

a angle ***A*** ________

b angle ***B*** ________

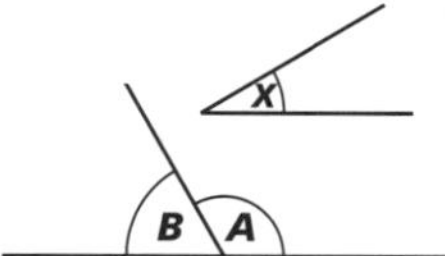

2 I fed 10 dogs either 6 or 7 biscuits each. I used 63 biscuits. How many dogs were given 7 biscuits? ________

3 Four equal angles make up a right angle. How big is each angle? ________

4 What number is halfway between:

a 87 and 99? ________ **b** 64 and 98? ________

5 ☐ ☐ ☐ represents 623.

☐ ☐ ☐ represents 850.

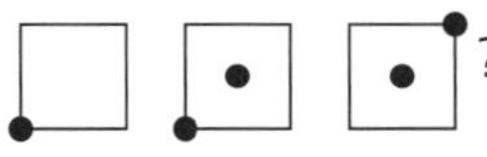

What number is represented by

☐ ☐ ☐? ________

6 What is half of 1 million? ________

Challenge

Use the subtraction from thousand strategy in 18:3 Question 1 to find:

a 5000 − 2537 *b* 8000 − 4291

19:1 out of 18

1. 3×7 ____
2. 6×8 ____
3. 4×4 ____
4. 8×4 ____
5. $\begin{array}{r} \$47.36 \\ +\ \$36.71 \\ \hline \end{array}$
6. $6 \times$ ____ $= 30$
7. 24 shared by 6. ____
8. 36 divided by 6. ____
9. $\frac{1}{2}$ of 16 ____
10. $\begin{array}{r} 6840 \\ -\ 4738 \\ \hline \end{array}$

11. 
Does this shape have rotational symmetry?

12. 2 weeks = ________ days

13. What is the 9 worth in 9465342? ________

14. 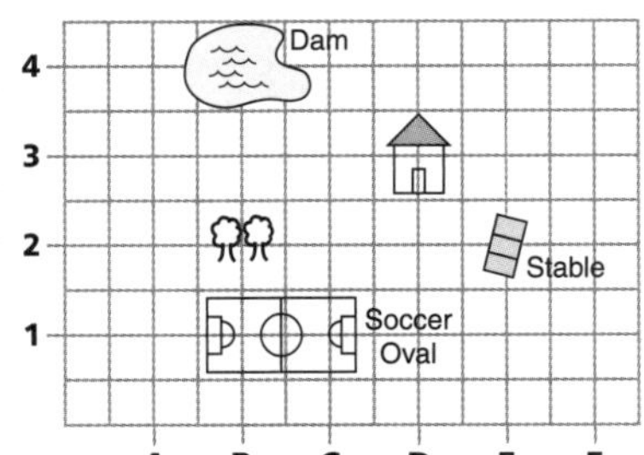

Write the coordinates of:

a (trees) ________ b (house) ________

What is located at:

c **B4**? ________ d **E2**? ________

15. List the factors of 12. ________
16. Round off 244 to the nearest hundred. ________
17. How many sides on 2 octagons? ________
18. 6, 12, 18, ____, ____, ____, ____, ____

19:2 out of 19

1. 9×5 ____
2. 5×8 ____
3. 9×8 ____
4. 7×6 ____
5. $\begin{array}{r} \$36.82 \\ +\ \$29.63 \\ \hline \end{array}$
6. 35 shared by 7. ____
7. $6 \times$ ____ $= 48$
8. $9 \times$ ____ $= 63$
9. 72 divided by 9. ____
10. $\begin{array}{r} 7642 \\ -\ 2586 \\ \hline \end{array}$

11. Circle the smallest decimal.
0·14 0·43 2·1 1·09

12. How many groups of 8 fish in 32? ________

13. Shade $\frac{2}{12}$ of this shape.

14. What is the order of rotational symmetry on this shape? ________

15. Write the numeral one million, nine hundred thousand, two hundred and forty-seven. ________

16. Round 7463372 to the nearest million. ________

17. $\frac{3}{8} + \frac{2}{8}$

18. **a** Estimate 145 + 68.
E = ____ + ____
= ____

b Estimate 614 − 293.
E = ____ − ____
= ____

19. Write the fraction shaded.
____

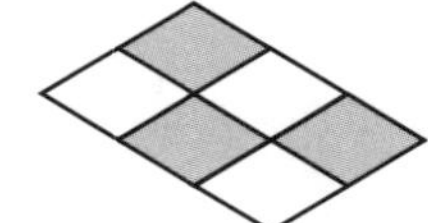

1. What object is:
   - **a** north of **A**? ____
   - **b** east of **A**? ____
   - **c** west of **A**? ____
   - **d** south of **A**? ____
2. Where do you end up if you start at S and move:
   - **a** 3 spaces east, 2 north, then 1 west? ____
   - **b** 5 spaces north, 5 east, then 4 south? ____
   - **c** 4 north, 3 south, then 5 east? ____
   - **d** 1 north, 5 east, then 3 west? ____

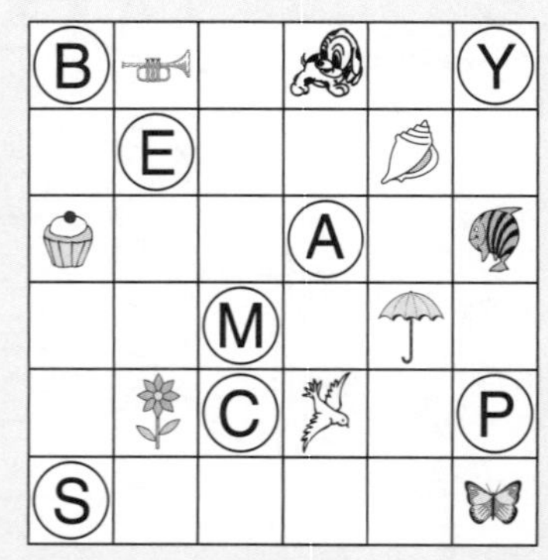

19:3 ☐ out of 11

1. 4000 − 750

2. 6000 − 250

3. 9000 − 2043

4. 2)82

5. 3)69

6. 6)72

7. Use the jump strategy for 933 − 271. ______

933

8.  What is the order of rotational symmetry on this shape? ______

9. I started at **X**, went 3 km east, then 3 km north, then travelled 3 km south east. Through which letter did I pass?

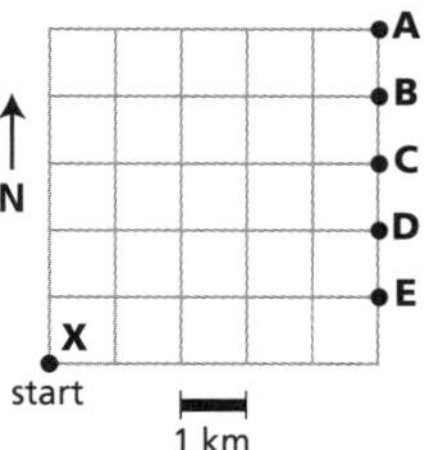

______

10. Write the numeral:

a 45 million ______

b 37 billion ______

11. 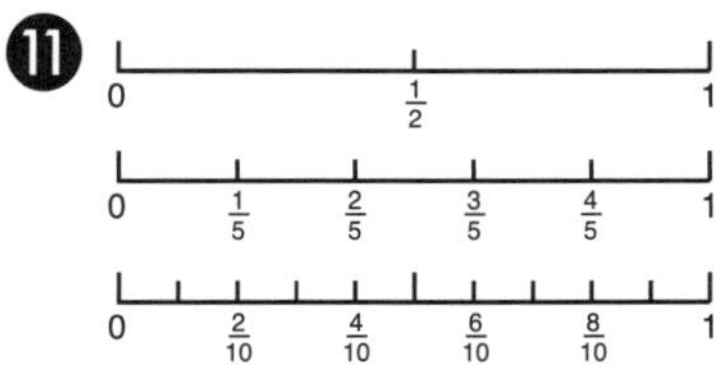

a $\frac{1}{2} = \frac{\square}{10}$ = ______ **b** $\frac{8}{10} = \frac{\square}{5}$ = ______

c $\frac{2}{5} = \frac{\square}{10}$ = ______ **d** $1 = \frac{\square}{5}$ = ______

19:4 Extension ☐ out of 8

1. We shared 5 oranges among 10 players. What would a fair share be for each? ______

2. Four groups of $1\frac{1}{2}$. ______

3. What is the least number of matches needed to make 12 triangles? ______

4. **a** ______ + 32 = 186

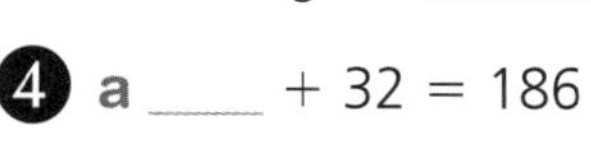

b 256 + ______ = 284

5. If # means 'add 15' and ^ means 'divide by 3', then the value of 9 ^ ^ # # # is ______.

6. If May 9th was a Monday, what day was May 30th? ______

7. At 8:30, the hour and minute hand make an angle on the clock. What is the size of the:

a acute angle? ______

b reflex angle? ______

8. 45 minutes after 3:45 pm. ______

Challenge

Draw and label a compass rose with N, NE, E, SE, S, SW, W and NW.

1 Which shape could be the result, if:

a I rotate the top heptagon? ______

b I reflect the top heptagon? ______

2 Which of the diagrams has an axis of symmetry? ______

3 How many diagonals has a heptagon? ______

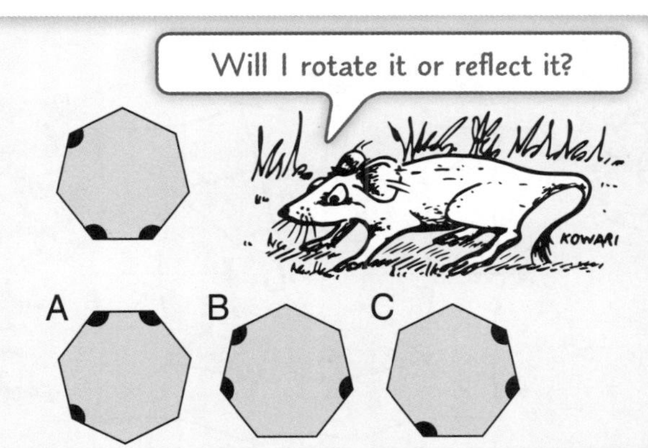

20:1 ☐ out of 15

1. 6×10 ______
2. 8×4 ______
3. 6×6 ______
4. 5×5 ______
5. $\begin{array}{r} 2846 \\ +\,4679 \\ \hline \end{array}$
6. $5 \times$ ______ $= 50$
7. 30 divided by 6. ______
8. 40 shared by 5. ______
9. $\frac{1}{2}$ of 26. ______
10. $\begin{array}{r} \$42.93 \\ -\,\$37.87 \\ \hline \end{array}$
11. Complete the following:

    216·93 = ______ hundreds
    ______ tens
    ______ units
    ______ tenths
    ______ hundredths

12. **A**  **B** 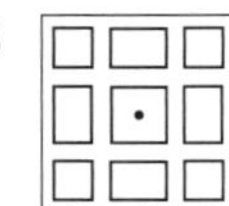 **C** 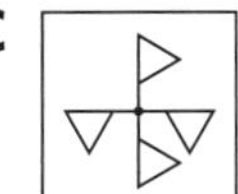

    a Which of these patterns does not have turning symmetry? ______
    b How many times does a tracing of pattern **A** exactly match the original as it turns through a full turn?
13. 5 red and 4 green marbles are in a bowl. Which colour is more likely to be chosen at random? ______

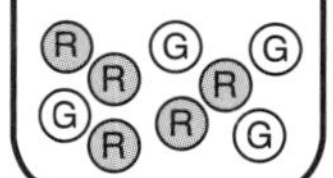

14. Write the smallest 7-digit number you can using only the digits 5, 6, 3 and 8. ______
15. Write the factors of 24. ______

20:2 ☐ out of 18

1. 9×7 ______
2. 8×8 ______
3. 5×7 ______
4. 4×9 ______
5. $\begin{array}{r} 6359 \\ +\,1773 \\ \hline \end{array}$
6. 9 squared. ______
7. 3 squared. ______
8. $49 \div 7$ ______
9. $64 \div 8$ ______
10. $\begin{array}{r} \$35.76 \\ -\,\$18.98 \\ \hline \end{array}$
11. $3\overline{)309}$
12. $7\overline{)84}$
13. $4\overline{)128}$
14. **A**  **B**  **C** 

    a Do all of these designs have rotational symmetry? ______
    b Which has rotational symmetry of order 10? ______
15. **a** 21 days = ____ weeks
    b 180 minutes = ____ hours
16. Round 4637 to the nearest:
    a ten ______ **b** thousand ______
17. **a** Kilometres in 5000 m ______
    b Litres in 3000 mL ______
    c Centimetres in 400 mm ______
    d Kilograms in 3000 g ______
18. Write as a decimal:
    a $\frac{3}{10}$ ______ **b** $\frac{7}{10}$ ______
    c $\frac{1}{2}$ ______ **d** $1\frac{1}{10}$ ______

20:3 out of 14

1. 7351 − 2672
2. 5000 − 370
3. 2000 − 1821
4. 2)42
5. 4)88
6. 7)91
7. 3)369
8. 2)846
9. 5)755
10. A B C

   a Which of these designs does not have rotational symmetry? ______
   b Which has rotational symmetry of order 5? ______
11. Is it equally likely that the spinner will land on 1, 2 or 3? ______

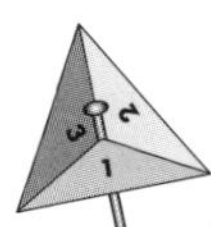

12. Cody is twice as old as Bill. If Cody is 30, how old is Bill? ______
13. Complete the following:

    434·58 = ______ hundreds
    ______ tens
    ______ units
    ______ tenths
    ______ hundredths

14. a Degrees in a right angle. ______
    b Degrees in a straight angle. ______
    c Degrees in a revolution. ______

20:4 out of 7

Extension

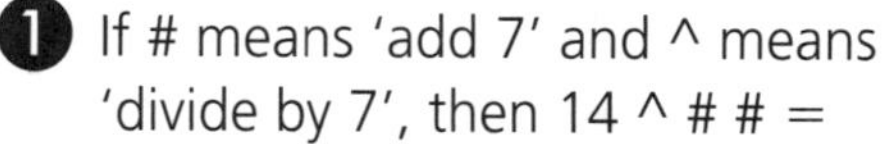

1. If # means 'add 7' and ^ means 'divide by 7', then 14 ^ # # = ______
2. I tossed 2 coins 40 times. This tally shows the results.

2 heads	𝍸 𝍸
2 tails	𝍸 𝍸
head and tail	𝍸 𝍸 𝍸 𝍸

   What fraction of the time did I get:
   a a head and a tail? ______
   b 2 heads? ______
3. If it is 5 March 2024 today, what will the date be in four weeks time? ______
4. I keep rabbits and birds. There are 5 heads and and 16 feet. How many birds do I have? ______

5. A zoo wants to buy 10 animals: lions and tigers. There must be at least one of each. How many different possible choices could they make? ______
6. 8000 − 3921 = 7999 + 1 − 3921 = ______
7. 18 + 23 + 32 + 7 + 5 ______

Challenge

Complete a design using rotational symmetry.

Chance

Is this game fair? ______

Two coins are tossed.

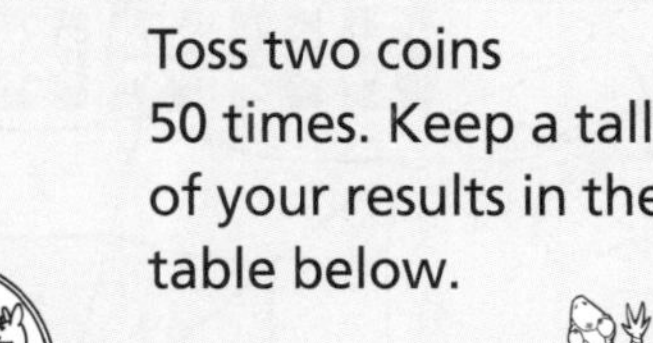

Toss two coins 50 times. Keep a tally of your results in the table below.

H and H	
T and T	
H and T	

21:1 ☐ out of 19

1. 10 × 10 ____
2. 34 × 10 ____
3. 64 × 100 ____
4. 53 × 10 ____
5. 5620 + 4687
6. 7 tens. ____
7. 23 hundreds. ____
8. 5 × 6 tens ____
9. 7 × 3 hundreds ____
10. $56.68 − $36.79

11. 10 students had to buy 6 books for high school. How many books were bought? ____

12. a Estimate 189 + 123.
    E = ____ + ____
    = ____

    b Estimate 823 − 319.
    E = ____ − ____
    = ____

13. a $\frac{1}{2}$ of one kilogram. ____
    b $\frac{1}{4}$ of one litre. ____

14. Is it equally likely that the number rolled on a 6-sided die will be 1, 2, 3, 4, 5 or 6? ____

15. Write the improper fraction for the shaded part.
    a 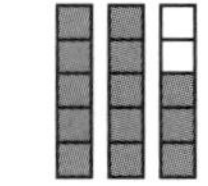 ____
    b 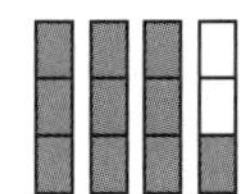 ____

16. Which is smaller, 2·71 or 2·37? ____

17. Write the numeral 37 million. ____

18. How many 50c apples can you buy for $4? ____

19. Write $\frac{17}{3}$ as a mixed number. ☐

21:2 ☐ out of 18

1. 58 × 100 ____
2. 28 × 100 ____
3. 7 × 70 ____
4. 7 × 60 ____
5. 7564 + 1636
6. 7 × 2 × 10 ____
7. 6 × 3 × 10 ____
8. 8 multiplied by $70. ____
9. 900 × 7 ____
10. $67.48 − $13.54

11. *To find the average, find the total length and divide by the number of fish.* The fish were 25 cm, 18 cm, 30 cm and 27 cm long. What was the average length of the fish? ____

12. Lydia wants to save $90 in 10 weeks. How much will she need to save each week? ____

13. Is each angle acute or obtuse?
    a 78° ____ b 96° ____

14. The pet shop was selling 145 budgerigars and 157 finches. How many budgerigars and finches were there altogether? ____

15. a 5700 mL = ____ L b 8451 g = ____ kg
    c 67 mm = ____ cm d 375 cm = ____ m

16. Write the numeral 46 billion. ____

17. a Complete this pattern.
    $\frac{22}{100}$, $\frac{21}{100}$, $\frac{20}{100}$, ____, ____, ____
    b The rule is ____.

18. Find the perimeter of a rectangle with side lengths 4 cm and 3 cm. ____

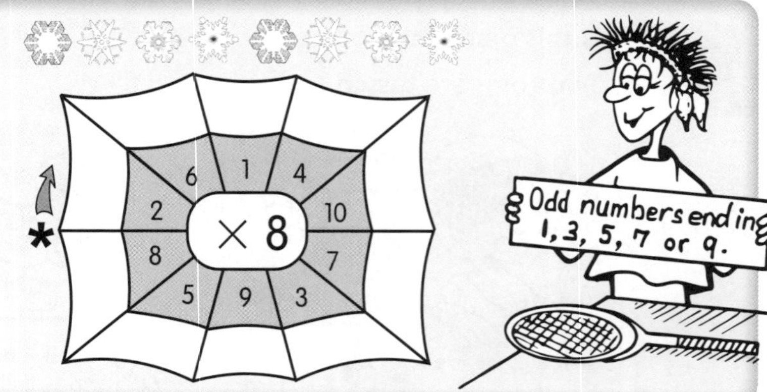

 ISBN 978 0 6557 0912 1

21:3 ☐ out of 18

1. $\begin{array}{r} 4325 \\ -\ 1782 \\ \hline \end{array}$

2. $\begin{array}{r} 5000 \\ -\ 2185 \\ \hline \end{array}$

3. $\begin{array}{r} 6308 \\ -\ 4992 \\ \hline \end{array}$

4. 3)28 r

5. 9)47 r

6. 7)31 r

7. 3)906

8. 2)648

9. 5)650

10. Estimate 439 + 35.
    E = ____ + ____
    = ____

11. Estimate 475 − 233.
    E = ____ − ____
    = ____

12. Elizabeth and Max throw a standard dice.
    Elizabeth wins if the number rolled is even. Max wins if the number rolled is 1 or 3. Is this a fair game? ____

13. The temperature this week was 24°C on Monday, 21°C on Tuesday and 27°C on Wednesday. What was the average temperature? ____

14. There are 24 melons in 1 case. How many melons are in 10 cases? ____

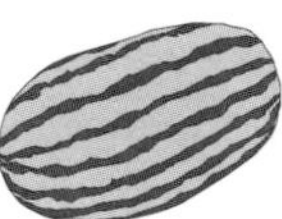

15. a Millimetres in 2·7 cm ____
    b Grams in 3·5 kg ____
    c Millilitres in 4·876 L ____

16. Write the fraction and percentage for each.

0·40	$\frac{\ }{100}$	%

0·95	$\frac{\ }{100}$	%

17. How many quarters in 3 apples? ____

18. How many groups of 5 cats in 25? ____

21:4 Extension ☐ out of 6

1. Our heights were 145 cm, 152 cm, 142 cm, 135 cm and 156 cm. What is the average height? ____

2. I opened a book. The product of the two page numbers I saw was 110. What were the numbers? ____

3. There are 8 petals on each flower.
   a How many petals altogether? ____
   b How many flowers would make 200 petals? ____

4. a 157 ÷ 3 ____ r ____
   b 157 ÷ 5 ____ r ____

5. 9 o'clock, quarter past 12, half past 9, 3 o'clock.
   At which of the above times are the hour and minute hands at right angles?
   ____

6. a 3 × 3 × 90 ____
   b 2 × 2 × 50 ____
   c 4 × 4 × 200 ____
   d 5 × 5 × 40 ____

Challenge

Write your own number patterns using fractions and decimals and the rule for each.

____
____
____
____

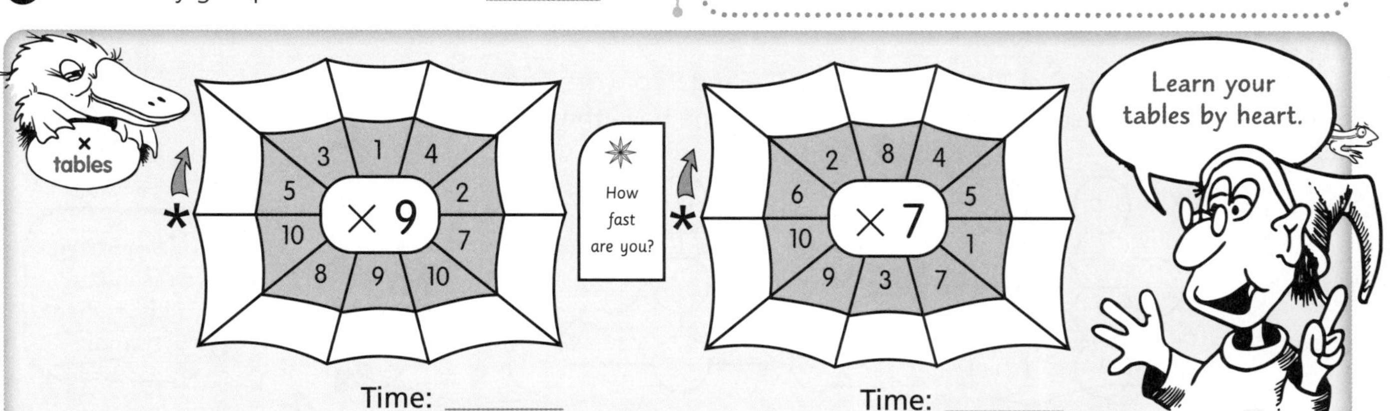

22:1 ☐ out of 20

1. 8×30 ____
2. 4×20 ____
3. 7×500 ____
4. 8×40 ____
5. $\begin{array}{r} 8354 \\ +\ 1397 \\ \hline \end{array}$
6. 81 hundreds. ____
7. 35 tens. ____
8. $8 \times$ ____ $= 40$
9. $9 \times$ ____ $= 36$
10. $\begin{array}{r} \$29.53 \\ -\ \$15.89 \\ \hline \end{array}$
11. Name the type of angle if its size is 163°. ____
12. Round 6375 to the nearest thousand. ____
13. $\frac{5}{8}$ means ____ out of ____.
14. What fraction of this shape is shaded? ____

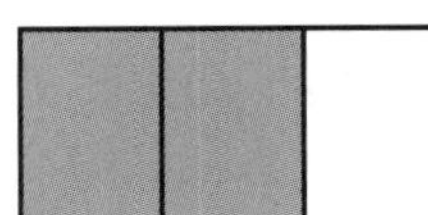

15. How many tens in 100? ____
16. How many tens could be taken from:
    a 476? ____ b 298? ____
17. Write an equivalent fraction for:
    a $\frac{1}{3}\frac{(\times 2)}{(\times 2)} = \frac{\square}{\square}$ b $\frac{4}{5}\frac{(\times 3)}{(\times 3)} = \frac{\square}{\square}$
18. Write the improper fraction for $5\frac{1}{2}$. $\frac{\square}{\square}$
19. $3\frac{1}{2}$, 4, $4\frac{1}{2}$, ____, ____, ____
20. 

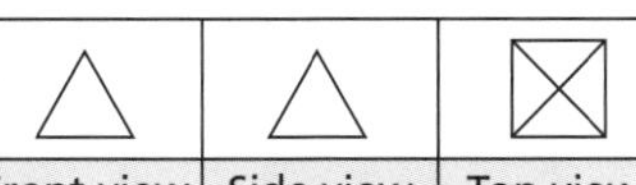

This shape is a ____.

22:2 ☐ out of 18

1. 8×50 ____
2. 9×600 ____
3. 7×700 ____
4. 6×600 ____
5. $\begin{array}{r} 3536 \\ +\ 3614 \\ \hline \end{array}$
6. 3×8 hundreds ____
7. 9×5 hundreds ____
8. 45 divided by 5. ____
9. 56 shared by 8. ____
10. $\begin{array}{r} \$99.46 \\ -\ \$53.87 \\ \hline \end{array}$
11. $1\frac{4}{10}$ as a decimal. ____
12. What was the average time taken to run a race if Annette took 13 seconds, Micah took 13 seconds, and Kwan took 10 seconds? ____
13. Order from smallest to largest.
    $\frac{15}{100}$, $\frac{21}{100}$, $\frac{11}{100}$,

14. Luke has 12 flowers in his garden. This is 4 times as many as Jo.
    How many has Jo? ____

15. 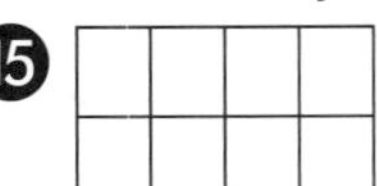
    a $\frac{5}{8} - \frac{3}{8} =$ ____ b $\frac{7}{8} - \frac{2}{8} =$ ____
    c $\frac{8}{8} - \frac{1}{8} =$ ____ d $\frac{6}{8} - \frac{5}{8} =$ ____
16. Write the mixed number for $\frac{14}{5}$. $\frac{\square}{\square}$
17. a $\frac{1}{2} = \frac{\square}{12}$ b $\frac{1}{4} = \frac{\square}{12}$

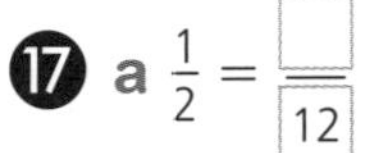

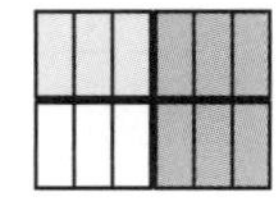

18. This is the net of a ____.

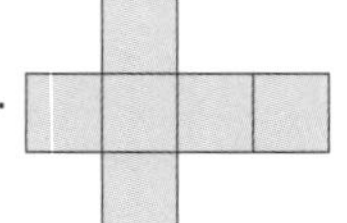

Strategy Time

In each case, use the digits 3, 4, 5, 7, 8 and 9 to write one digit in each circle so that the numbers in each line have the same total.

a

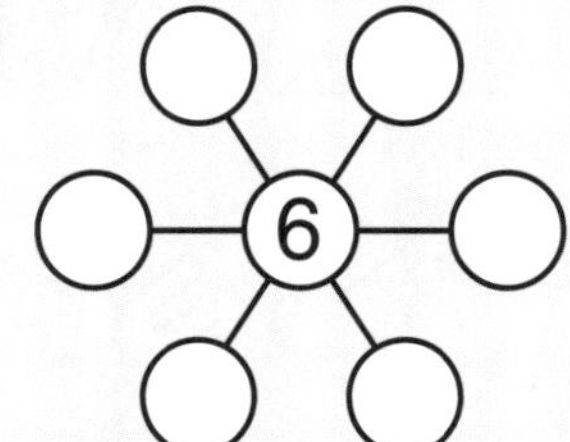

b

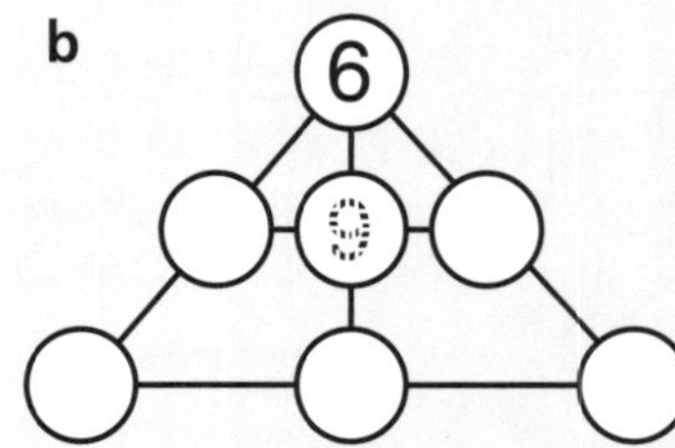

 • *AUSTRALIAN SIGNPOST MATHS NSW 5 MENTALS* • ISBN 978 0 6557 0912 1

22:3 out of 16

1. $\begin{array}{r} 6312 \\ -\ 943 \\ \hline \end{array}$

2. $\begin{array}{r} 4431 \\ -\ 653 \\ \hline \end{array}$

3. $\begin{array}{r} 9013 \\ -\ 2437 \\ \hline \end{array}$

4. $5\overline{)14}$ r

5. $3\overline{)19}$ r

6. $4\overline{)368}$

7. $3\overline{)360}$

8. $5\overline{)913}$

9. $6\overline{)378}$

10. Draw the front, side and top elevation.

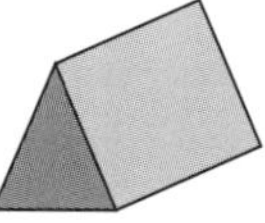

Front

Side

Top

11. Write the improper fraction for $3\frac{4}{6}$.

12. Plot this set of points and join them in order.
1A, 1B, 0B, 1D,
2D, 3B, 2B, 2A,1A.

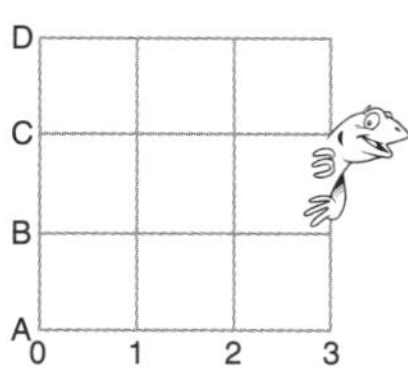

13. 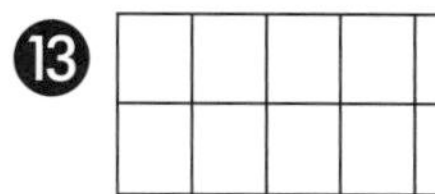

a $\frac{8}{10} - \frac{4}{10} =$ ______ b $\frac{9}{10} - \frac{7}{10} =$ ______

14. Write the mixed number for $\frac{14}{5}$.

15. a $\frac{1}{2} = \frac{\square}{12}$ b $\frac{1}{4} = \frac{\square}{12}$

16. a 9 km = ______ m b 4 cm = ______ mm

22:4 Extension out of 4

1. Which part is the:
   a numerator? ______
   b denominator? ______
   c Which of the fractions shaded would have the smaller denominator? ______

$\frac{7}{10}$

A B

2. a 0·7 ÷ 7 ______
   b 0·7 ÷ 10 ______
   c 0·7 ÷ 2 ______
   d 0·7 ÷ 5 ______

0·7 = 0·70

3. a 2·5, 3·6, 4·7, ______, ______, ______, ______
   b 0·8, 1·0, 1·2, ______, ______, ______, ______

4. Danni hit 24 runs. Maddie hit 4 times as many. How many runs did they hit altogether? ______

Challenge

Draw a design that has rotational symmetry.

23:1 ☐ out of 16

1. 57 + 59 ______
2. 157 + 12 ______
3. 146 − 21 ______
4. 3 × 4 ______
5. $\begin{array}{r} 2647 \\ +\,7253 \\ \hline \end{array}$
6. 4 times 4. ______
7. 8 groups of 5. ______
8. 5 × ______ = 10
9. 3 × ______ = 18
10. $\begin{array}{r} \$56.37 \\ -\,\$35.74 \\ \hline \end{array}$
11. Write the improper fraction for $7\frac{1}{2}$. $\frac{\square}{\square}$
12. Write an equivalent fraction for:

    a $\frac{1}{5}\frac{(\times 3)}{(\times 3)} = \frac{\square}{\square}$ b $\frac{1}{3}\frac{(\times 4)}{(\times 4)} = \frac{\square}{\square}$

13. 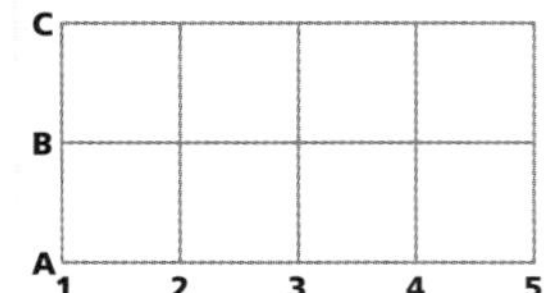

    1A, 3C, 4C, 5A, 1A

    Plot this set of points. Join each one to the next. Then name the shape you have drawn.

    ______

14. This is the net of which shape?

    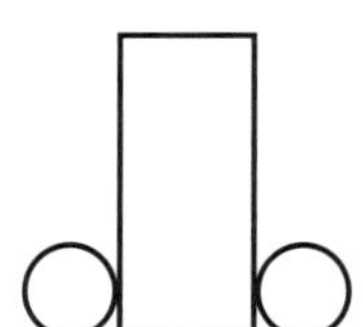

    ______

15. 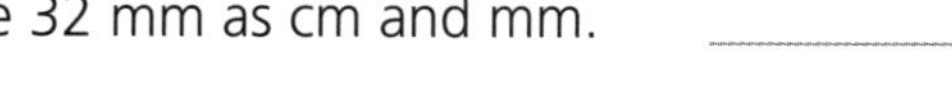

    Estimate then measure the length of this bar, correct to the nearest millimetre.

    Estimate = ______ Measure = ______

16. Write 32 mm as cm and mm. ______

23:2 ☐ out of 15

1. 9 × 8 ______
2. 6 × 5 ______
3. 5 × 7 ______
4. 9 × 6 ______
5. 5 × ______ = 30
6. 28 divided by 7. ______
7. Add 259 to 247. ______
8. 359 plus 356. ______
9. $10\overline{)437}$
10. $7\overline{)816}$
11. $4\overline{)625}$
12. Write the mixed number for $\frac{7}{2}$. $\frac{\square}{\square}$
13. Of which solid is this a net?

    ______

14. 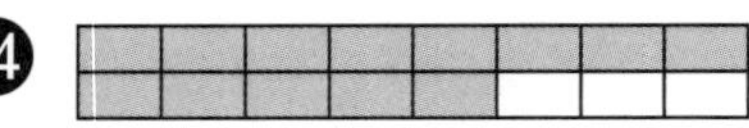

    a $\frac{3}{4} = \frac{\square}{20}$ b $\frac{1}{2} = \frac{\square}{20}$

    c $\frac{2}{20} = \frac{\square}{10}$ d $\frac{10}{20} = \frac{\square}{10}$

15. 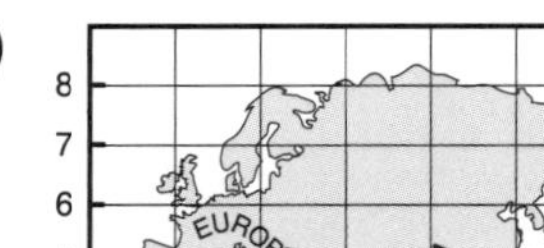

    a What continent is north-west of Australia? ______

    What continent has coordinates:

    b A4? ______

    c J3? ______

    d What coordinates could be used to locate Australia? ______

Turn to ID card C on page 8.
Give the answers for these numbers.

(5)	______	(6)	______
(7)	______	(8)	______
(9)	______	(10)	______
(11)	______	(12)	______
(13)	______ shapes	(14)	______ shapes

Make a study card:
Put questions on one side and answers on the other.

23:3 out of 13

1. 5971 + 1435
2. 6984 + 1132
3. 7823 + 1697
4. 3)26 r
5. 5)39 r
6. 6)52 r
7. Write the improper fraction for $8\frac{7}{9}$.
8. Is $\frac{3}{5}$ equal to $\frac{9}{10}$? ______
9. Which prism has this net? ______

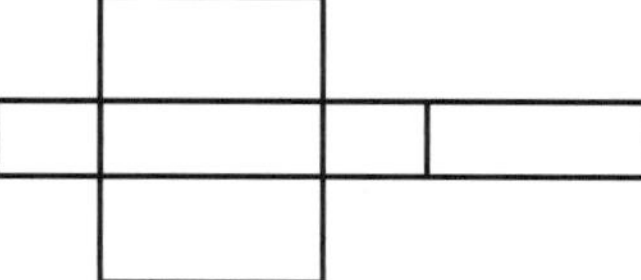

10. Write the mixed number for $\frac{15}{2}$.
11. Multiply the numerator and denominator by 2 to find an equivalent fraction

    a $\frac{1}{3} = \frac{\square}{\square}$ **b** $\frac{4}{5} = \frac{\square}{\square}$

12. 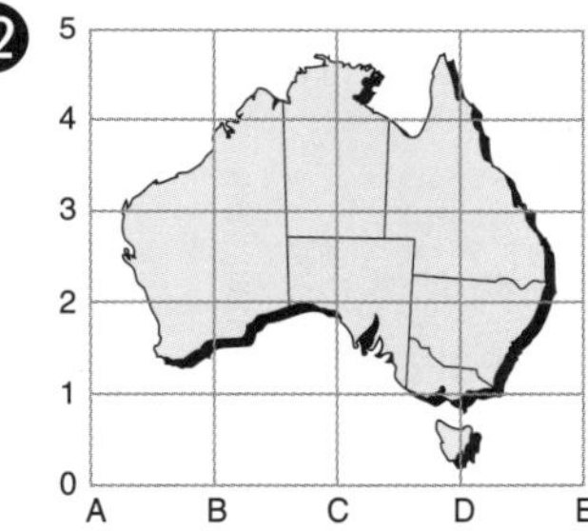

One grid reference for Queensland is D4. Give the best possible reference for:

a Qld ______ **b** NSW ______
c SA ______ **d** Vic ______

13. Write the temperature shown.

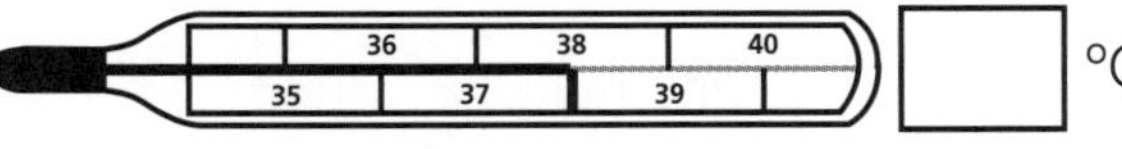

☐ °C

23:4 Extension out of 5

1. 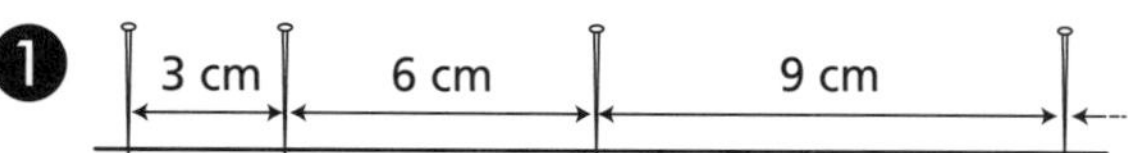

Pins were stuck onto a board using the pattern shown above. What was the distance between:

a the 5th and 6th pins? ______
b the 1st and 6th pins? ______

2. 118 mm + 115 mm + 22 mm + 45 mm ______
3. Write as millimetres:

   a 5·85 m ______
   b 16·4 m ______

4. The coordinates of the 4th corner if:

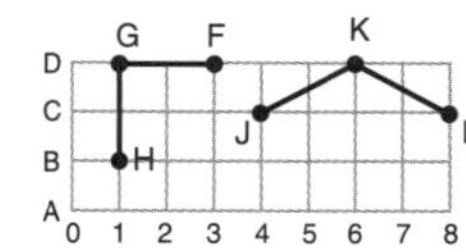

a **F**, **G**, **H** is to be a square. ______
b **J**, **K**, **L** is to be a rhombus. ______

5. **a** $2\frac{1}{2}$m = ______ cm **b** $2\frac{1}{2}$cm = ______ mm

Challenge

Choose a household object.
Draw and label the top, side and front view below.

Each student traced a path on this grid.

1. Wayne started at C, moved 2 up, 6 right, 1 down, then 5 left. Where did he end up? ______
2. Chloe ended up at H after moving 3 up, 5 right, 3 down, 3 left then 2 up. Where did she start from? ______

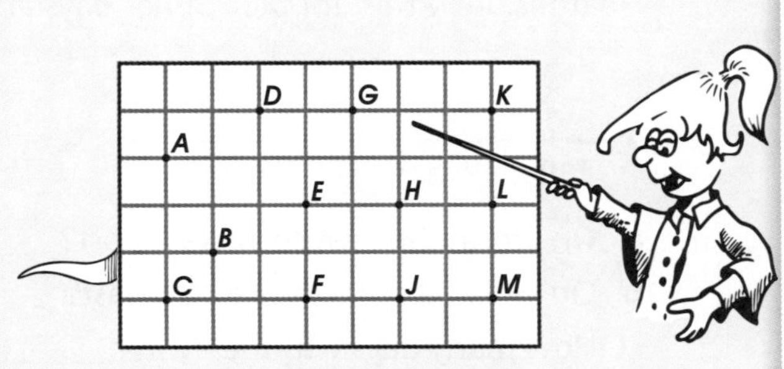

24:1 ☐ out of 22

1. 137 + 42 ______
2. 99 − 54 ______
3. 4 × 7 ______
4. 6 × 4 ______
5. 3 groups of 9. ______
6. Multiply 45 by 10. ______
7. 80 divided by 8. ______
8. 100 shared by 10. ______
9. $10\overline{)360}$
10. $7\overline{)274}$
11. $8\overline{)385}$
12. Order these decimals from smallest to largest.
    5·7   5·03   5·36   5·39
    ______
13. Write an equivalent fraction for:
    a $\frac{1}{2}\frac{(\times 5)}{(\times 5)} = \frac{\square}{\square}$ **b** $\frac{1}{6}\frac{(\times 3)}{(\times 3)} = \frac{\square}{\square}$

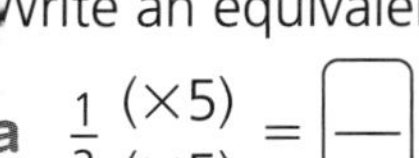

14. Does 10 × 0·1 = 1? ______
15. Round 4·84 to 1 decimal place. ______
16. Millimetres in 1 centimetre. ______
17. Write the number 35 thousand. ______
18. Would you use km, m, cm or mm to measure:
    a the length of a soccer field? ______
    b the width of a pencil? ______
19. Use the decimal form to write 39 mm. ______ cm
20. Write 6·56 m as centimetres. ______
21. Use am or pm to write 08:27. ______
22. The distance around a shape is called the p ______.

24:2 ☐ out of 19

1. 10 × 0·01 ______
2. 4 × 5 ______
3. 89 + 45 ______
4. 98 + 67 ______
5. 500 × 7 ______
6. 100 × 0·02 ______
7. 198 plus 576. ______
8. 2 squared. ______
9. $7\overline{)375}$
10. $9\overline{)386}$
11. $10\overline{)274}$
12.

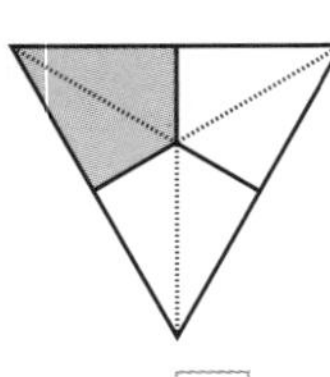

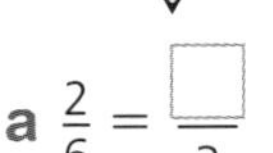

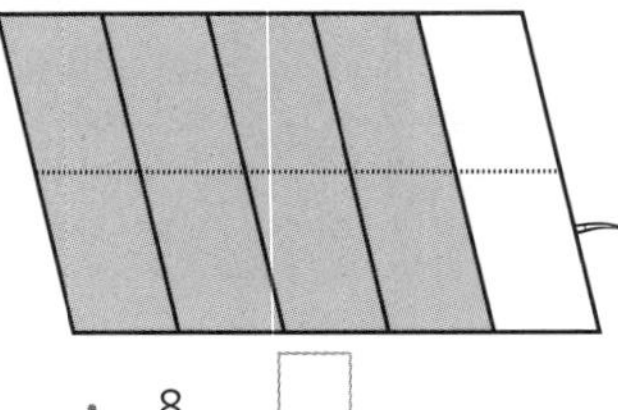

a $\frac{2}{6} = \frac{\square}{3}$ **b** $\frac{8}{10} = \frac{\square}{5}$

13. Order these decimals from smallest to largest.
    3·06 m   3·6 m   3·58 m   3·68 m
    ______
14. Use decimal form to write these as metres:
    a 385 cm ______   **b** 836 cm ______
15. Does 10 × 0·01 = 0·1? ______
16. **a** Round 7·38 to 1 decimal place. ______
    b Round 8·296 to 2 decimal places. ______
17.

Estimate then measure the length of this bar in millimetres.
Estimate = ______ mm   Measure = ______ mm

18. Perimeter of a square of side length.
    a 8 m ______   **b** 4·5 m ______
19. The difference in time between 5:35 am and 8:14 am. ______

Using a dot plot

Heather drew this dot plot of her days absent this year.

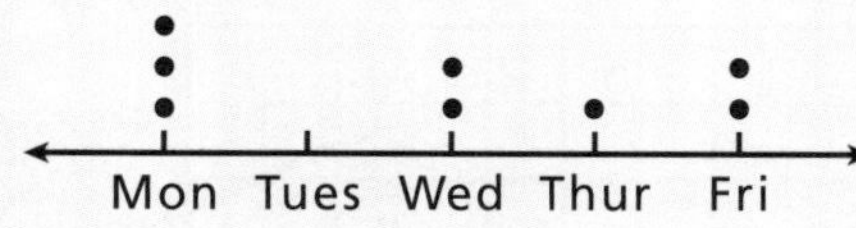

a On which day was she away most? ______
b On which day was she away least? ______
c How many days was she away? ______

d Draw a dot plot of these results.
1, 3, 2, 2, 1, 2, 2, 0, 1, 1, 2, 1, 3, 0, 2.

0 1 2 3

24:3 out of 12

1. 9)756
2. 6)375
3. 10)742
4. Order these decimals from smallest to largest.
   26·09 L   26·59 L   26·8 L   31 L   6·99 L
5. Does 10 × 0·01 = 1?
6. a Round 3·75 to 1 decimal place.
   b Round 3·057 to 2 decimal places.
7. 

   a Place a dot at 9·236 and 9·253.
   b Write the number that is halfway between 9·24 and 9·25.
8. The length of this pencil.
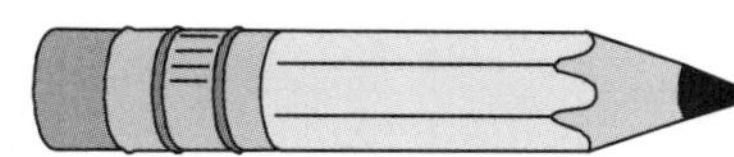
   Estimate = ______ mm
   Measure = ______ mm
9. 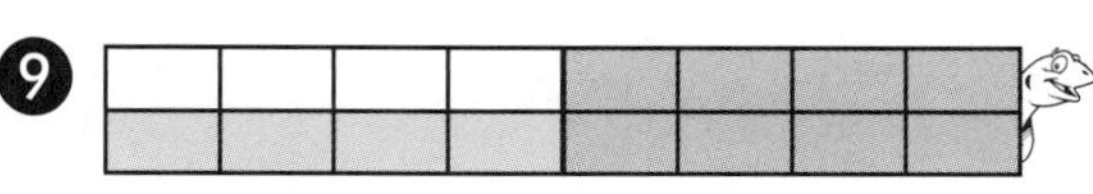
   a $\frac{4}{16} = \frac{\square}{4}$ b $\frac{8}{16} = \frac{\square}{2}$
   c $\frac{6}{16} = \frac{\square}{8}$ d $1 = \frac{\square}{16}$
10. Use decimal form to write these as centimetres.
    a 57 mm ______   b 39 mm ______
11. Write 3 cm 6 mm as millimetres.
12. Use am or pm to write 16:45.

24:4 Extension out of 8

1. What is the perimeter of a square of area 4 cm^2?
2. 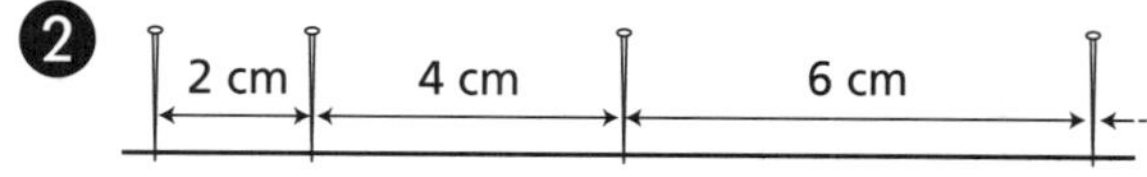

   Pins were stuck onto a board using the pattern shown above. What was the distance between:
   a the 5th and 6th pins?
   b the 1st and 6th pins?
3. Jill's skipping rope was 145 cm long. She cut 3·4 cm off. How long is it now?
4. 250 metres is ______ of a kilometre.
5. Estimate, then measure your height.
   Estimate = ______   Measure = ______
6. The difference in time between:
   a 10:39 am and 11:14 pm
   b 03:56 and 19:32
7. What is a quarter of $237?
8. a 3, 12, 21, ___, ___, ___, ___
   b 5, 24, 43, ___, ___, ___, ___
   c 142, 119, 96, ___, ___, ___, ___

Challenge

Write equivalent fractions for:

a $\frac{1}{2}$ b $\frac{1}{3}$ c $\frac{1}{4}$

24-hour time

midnight	3 am	6 am	9 am	noon	3 pm	6 pm	9 pm	midnight
00:00	03:00	06:00	09:00	12:00	15:00	18:00	21:00	24:00

Add 12 for pm.

Write the 24-hour time for each of these digital times.

a 6 pm ______   b 9:30 am ______
c 7 am ______   d 10:30 pm ______
e noon ______   f midnight ______ or ______

25:1 out of 19

1. 167 + 9 ______
2. 5 × 30 ______
3. 16 ÷ 4 ______
4. 24 ÷ 4 ______
5. 35 divided by 5. ______
6. Multiply 7 by 4. ______
7. Add 206 to 143. ______
8. 150 plus 150. ______
9. $10\overline{)385}$
10. $2\overline{)804}$
11. $6\overline{)704}$
12. Order these decimals from smallest to largest.

    2·08 2·9 2·46 2·74

    ______
13. Write the equivalent fraction for:

    a $\frac{3}{4}\frac{(\times 2)}{(\times 2)} = \frac{\square}{\square}$ b $\frac{2}{3}\frac{(\times 4)}{(\times 4)} = \frac{\square}{\square}$
14. a Round 7·38 to 1 decimal place. ______

    b Round 6·037 to 2 decimal places. ______
15. 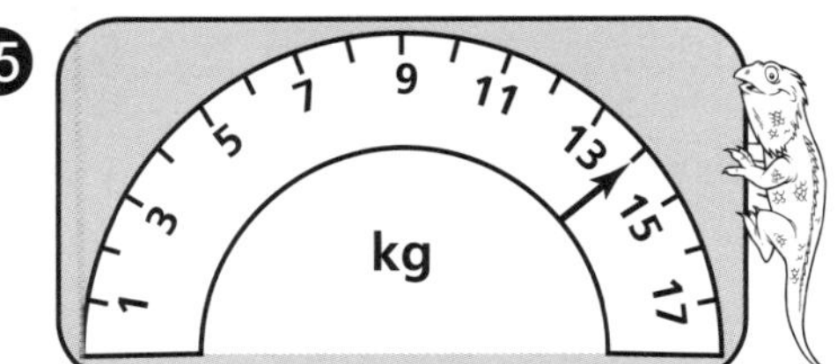

    The measure is ______ kg.
16. Is the length of a cricket bat closer to 30 cm, 1 m or 1 km? 

    ______
17. The difference in time between 3:45 pm and 4:15 pm. ______
18. Use 24-hour time to write 7:25 pm. ______
19. Use 12-hour time to write 16:47. ______

25:2 out of 17

1. 6 × 30 ______
2. 8 × 900 ______
3. 354 + 296 ______
4. 563 + 364 ______
5. 900 divided by 9. ______
6. Multiply 6 by 300. ______
7. 63 shared by 7. ______
8. 7·45 × 100 ______
9. $6\overline{)253}$
10. $5\overline{)379}$
11. $10\overline{)463}$
12. For this rectangle, find: 

    a the perimeter ______

    b the area ______
13. Order these decimals from smallest to largest.

    7·355 17·09 17·5 17·3

    ______
14. a Round 7·94 to 1 decimal place. ______

    b Round 3·058 to 2 decimal places. ______
15.  

    length = 82·125 cm length = 29·4 cm

    Write the length, to the nearest cm, of the:

    a lizard ______

    b echidna ______
16. This container is holding ______ mL. 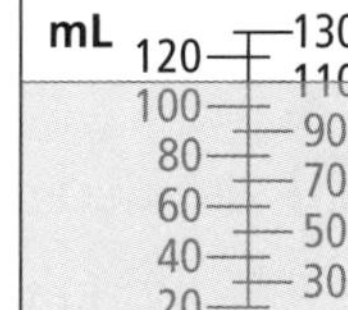

17. Use 24-hour time to write 2:46 am. ______

Using a dot plot

Luke drew this dot plot of the number of goals kicked in each game.

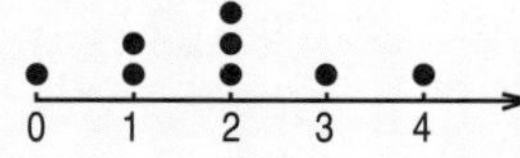

a What number was kicked most often? ______

b How many times was one goal kicked? ______

c How many goals were kicked altogether? ______

A, C, A, D, B, B, B, A, A, C

d Draw a dot plot of these results.

25:3 ☐ out of 17

1. $10\overline{)856}$ 2. $2\overline{)408}$ 3. $10\overline{)463}$
4. Would we use kilograms, grams or tonnes to measure the mass of a truck? ______
5. Order these decimals from smallest to largest.

   10·096 10·73 10·384 10·323

   ______
6. Does 100 × 0·01 = 0·1? ______
7. **a** Round 4·11 to 1 decimal place. ______

   b Round 8·494 to 2 decimal places. ______
8. 

   length = 42·93 cm

   length = 60·576 cm

   Write the length, to the nearest cm, of the:

   a platypus ______ **b** koala ______
9. 4546 m = ______ km
10. The value of the 6 in 5 364 718. ______
11. Use 24-hour time to write 3:56 pm. ______
12. Use 12-hour time to write 19:41. ______
13. Circle the numbers divisible by 2.

    37 16 8 13 1 4
14. Write all the factors of 24. ______
15. Write 4656 g as kg and g. ______
16. How many grams in 8·5 kg. ______
17. True or false: 6·5 kg > 6589 g ______

25:4 Extension ☐ out of 8

1. ☐ and △ are counting numbers less than 10 and ☐ + △ +7 = 24.

   So ☐ and △ are ______ and ______.
2. Complete the pattern below.

   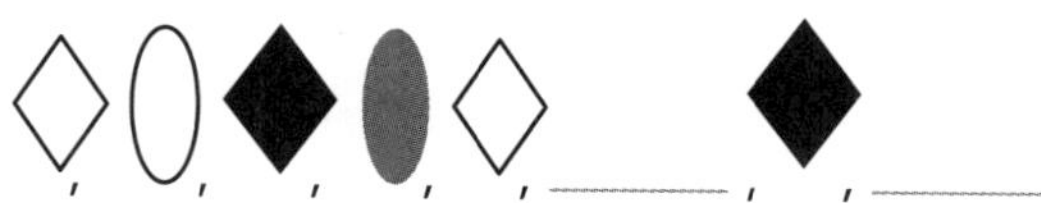

   ◇, ⬯, ◆, ⬮, ◇, ______, ◆, ______
3. If the pattern above continued, what would the 56th shape be? ______
4. Sarnya used 890 L to fill 10 identical buckets. How much was in each bucket if it was shared equally. ______
5. Which of the numbers 68, 198, 315 and 455 are divisible by:

   a 2 and 3? ______ **b** 5 and 9? ______
6. Write all the factors of 100. ______
7. How many days are in the last 6 months of the year? ______
8. 193, 275, 357, ______, ______, ______

Challenge

List objects that would be measured in:

g	kg	t

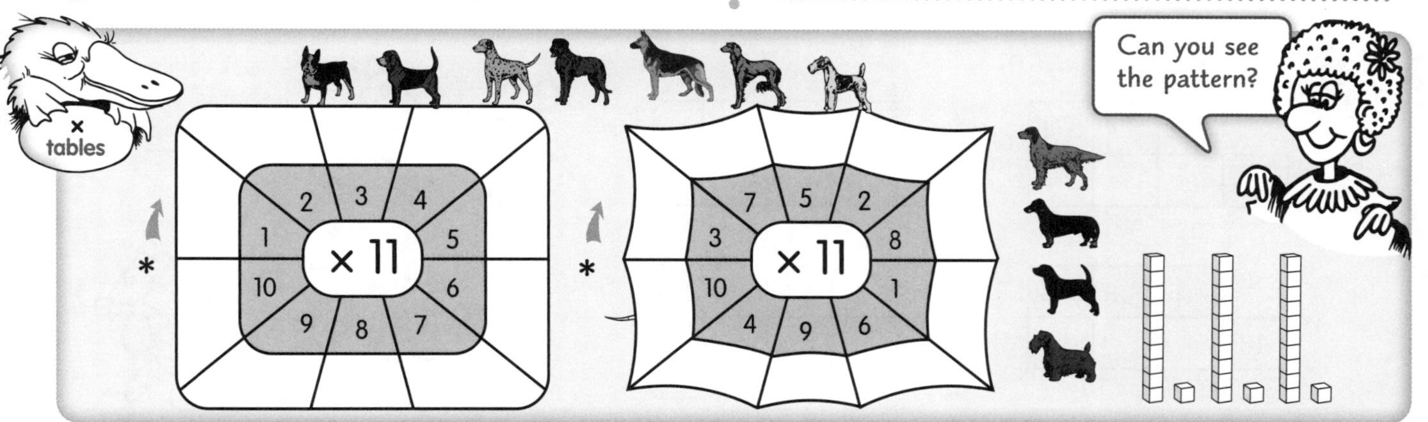

26:1

[ ] out of 21

1. 10 × 25 ______
2. 100 × 18 ______
3. 5 × 30 ______
4. 9 × 2 ______
5. 10 groups of 7. ______
6. 3 multiplied by 5. ______
7. 8 divided by 2. ______
8. 8 shared by 4. ______
9. $7\overline{)705}$
10. $10\overline{)745}$
11. $3\overline{)303}$
12. Does 100 × 0·01 = 1? ______
13. Order these decimals from smallest to largest.

    7·04 8·34 7·27 8·09

    ______
14. Circle the numbers divisible by 3.

    12 3 8 6 4 15 7
15. 2 kg of fish cost $8.00.
    What is the cost of 1 kg of fish?

    ______
16. Write 5367 g as kilograms and grams. ______
17. Write all the factors of 28. ______
18. True or false: 7 kg > 6999 g? ______
19. What is the average of 6 and 10? ______
20. (Rectangle: 6 cm by 1 cm)

    In this rectangle, find:

    a the perimeter ______

    b the area ______
21. Write the decimal for 35 hundredths. ______

26:2

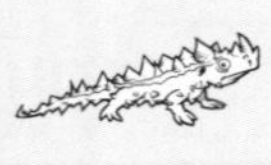

[ ] out of 22

1. 20 × 25 ______
2. 13 × 8 ______
3. 14 × 4 ______
4. 16 × 25 ______
5. (6 × 40) + (6 × 3) ______
6. (4 × 10) + (4 × 5) ______
7. (10 × 8) + (4 × 8) ______
8. (10 × 5) + (4 × 5) ______
9. $10\overline{)503}$
10. $2\overline{)816}$
11. $10\overline{)850}$
12. 3 × 63
    = (3 × 60) + (3 × 3)
    = ______
13. 6 × 37
    = (6 × 30) + (6 × 7)
    = ______
14. Write the first 6 multiples of 8.

    ______
15. Does 1000 × 0·01 = 10? ______
16. Order these decimals from smallest to largest.

    2·005 4·02 4·009 2·08

    ______
17. Write all the factors of 32. ______
18. a Grams in a $\frac{1}{2}$ kilogram. ______

    b Grams in $1\frac{1}{2}$ kilograms. ______
19. What is the average of 63 and 47? ______
20. True or false: 13 kg < 15 000 g ______
21. Look at the rectangle in 26:1, Question 20.
    If the dimensions were doubled, what would be:

    a the area? ______

    b the perimeter? ______
22. Order 4·65, 4·07, 4·109 and 4·804 from smallest to largest.

    ______

	÷ 2	÷ 4	÷ 1
8			

	÷ 2	÷ 12	÷ 6	÷ 3
12				

	÷ 6	÷ 9	÷ 3
18			

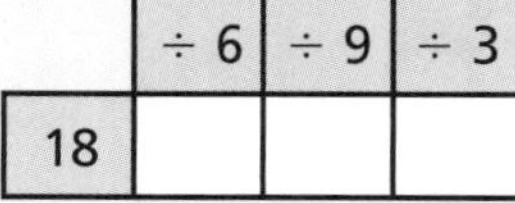

	÷ 2	÷ 6	÷ 4	÷ 8
24				

 • *AUSTRALIAN SIGNPOST MATHS NSW 5 MENTALS* • ISBN 978 0 6557 0912 1

26:3

out of 17

1. 18 × 4
   = ____ × 2 × 4
   = ____

2. 32 × 25
   = ____ × 4 × 25
   = ____

3. $10\overline{)648}$

4. $10\overline{)149}$

5. $10\overline{)915}$

6. Write all the factors of 36.
   ____

7. Write the first 4 multiples of 6. ____

8. Grams in $4\frac{1}{2}$ kg. ____

9. Order these decimals from smallest to largest.
   6·276   6·007   8·046   8·105
   ____

10. Rewrite 2967 g as kilograms and grams. ____

11. Circle the measure correctly written.
    A 3 cms **B** 3 cm **C** 3 CM **D** 3 Cm

12. The perimeter of the black rectangle is ____ centimetres.

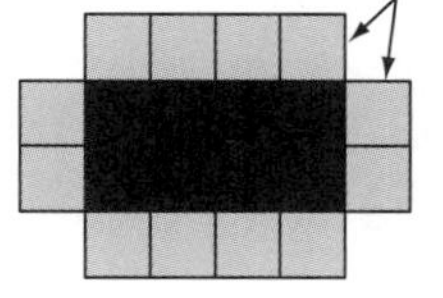

13. True or false: 2 kg < 2317 g? ____

14. What is the average of 48 and 56? ____

15. How many kilograms in 3000 g? ____

16. For a square that is 4 cm long, what is the:
    a perimeter? ____ **b** area? ____

17. For a square that is 8 cm long, what is the:
    a perimeter? ____ **b** area? ____

26:4

Extension

out of 8

1. The distance around each square is 16 metres. How far is it around the rectangle?

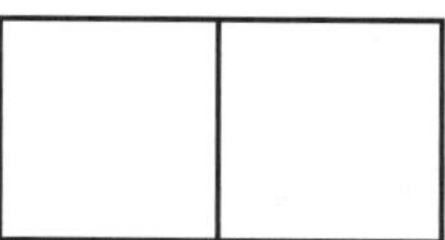

   ____

2. Which of the numbers 34, 95, 276 and 135 is divisible by both:
   a 2 and 3? ____ **b** 5 and 9? ____

3. The whole has a value of 24.
   Write the value of each shaded part.

   24

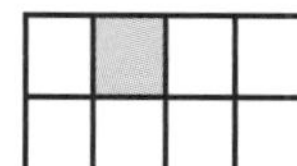

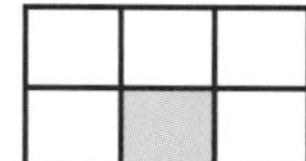

   ____ ____

4. What will the time on this watch be in 11 hours and 40 minutes?

   ____

5. **a** $2\frac{1}{2}$ m = ____ cm **b** $2\frac{1}{2}$ cm = ____ mm

6. What is the perimeter of a square with an area of 16 cm^2? ____

7. 357, 548, 739, ____, ____, ____, ____

8. What is a quarter of 5642? ____

List areas that would be measured in m^2 (e.g. a tennis court).

____

____

____

Factors of a number

Write all the factors of:

a 15 ____, ____, ____, ____

b 21 ____, ____, ____, ____

c 16 ____, ____, ____, ____, ____

d 12 ____, ____, ____, ____, ____, ____

If two numbers multiply to give 24, then they are factors of 24.

Now 24 = **1** × **24**, 24 = **2** × **12**,
   24 = **3** × **8**, 24 = **4** × **6**

The factors of 24 are 1, 24, 2, 12, 3, 8, 4 and 6.

 • *AUSTRALIAN SIGNPOST MATHS NSW 5 MENTALS* • ISBN 978 0 6557 0912 1

27:1 out of 20

1. 20 × 3 ______
2. 20 × 8 ______
3. 4 × 300 ______
4. 2 × 90 ______
5. $\begin{array}{r} 2745 \\ +\,6356 \\ \hline \end{array}$
6. 0·1 × 10 ______
7. 54 less than 100. ______
8. Halve 462. ______
9. Double 314. ______
10. $\begin{array}{r} \$86.85 \\ -\,\$56.07 \\ \hline \end{array}$
11. Write the first 6 multiples of 3.

    ____
12. What is the average of 13 and 17? ______
13. Use doubling to find:

    a 2 × 14 ______ **b** 4 × 14 ______
14. Complete the patterns below.

    a 8, 12, ______, ______, 24, 28, ______, ______

    b 7, 12, ______, 22, ______, 32, ______, ______
15. Subtract 104 from 659. ______
16. 4 × 27 = (4 × 20) + (4 × ______) = ______
17. Convert 56 mm to centimetres, using a decimal point. ______
18. **a** 5647 g = ______ kg

    b 8342 mL = ______ L
19. Write the numeral 47 million. ______
20. Draw a triangle with all acute angles on the left. Draw a scalene triangle on the right.

27:2 out of 18

1. 6 × 80 ______
2. 50 × 9 ______
3. 781 − 169 ______
4. 439 − 175 ______
5. 8 squared. ______
6. 0·8 × 100 ______
7. 365 plus 497. ______
8. 526 minus 215. ______
9. $5\overline{)375}$
10. $7\overline{)704}$
11. $3\overline{)735}$
12. 4 × 17 = (4 × 10) + (4 × 7) = ______
13. I earned $47 on Monday, $76 on Tuesday and $57 on Wednesday. What was my average earnings per day? ______
14. Use doubling to find:

    a 2 × 25 ______ **b** 4 × 25 ______
15. Write the first 6 multiples of 4.

    ____
16. Complete the pattern below.

    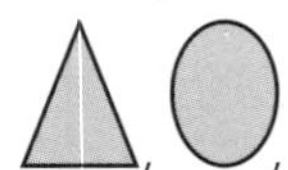, ______, ▲, ______, 
17. What would be the 33rd shape in this pattern? ______
18. 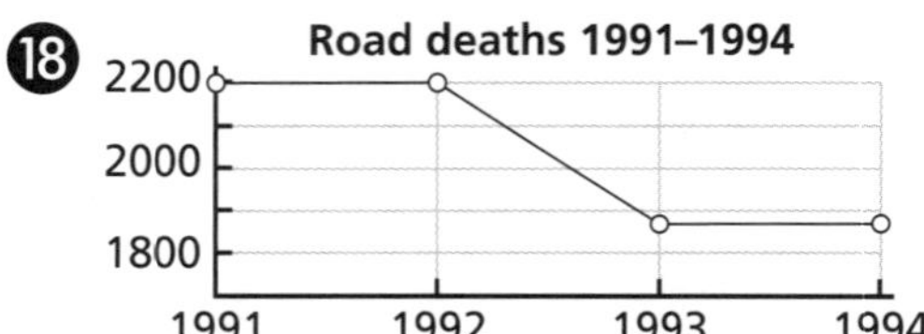

    a Estimate the total number of road deaths from 1991 to 1994. ______

    b In which year do you think breath tests for alcohol started? ______

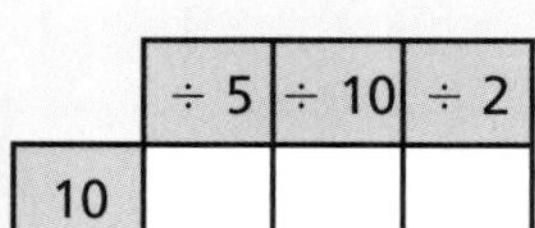

	÷ 5	÷ 10	÷ 2
10			

	÷ 3	÷ 1	÷ 9
27			

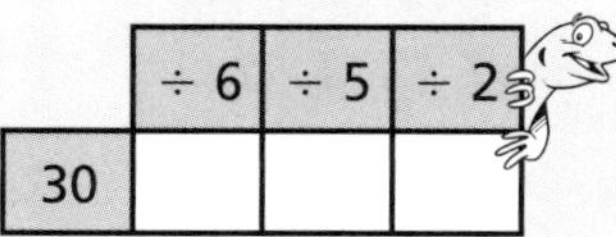

	÷ 6	÷ 5	÷ 2
30			

	÷ 2	÷ 6	÷ 7
42			

I have 27 lollies. How many people could be given 9 lollies?

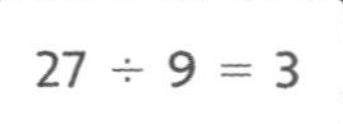

27:3 out of 11

1. 16 × 5
= ____ × 2 × 5
= ____

2. 16 × 25
= ____ × 4 × 25
= ____

3. $2\overline{)814}$

4. $2\overline{)806}$

5. $4\overline{)416}$

6. 5 × 54 = (5 × 50) + (5 × 4) = ____

7. Write the first 6 multiples of 9.
____

8. My friends and I scored 24 runs, 58 runs, 53 runs and 29 runs in cricket. What was our average score? ____

9. Complete this pattern.

1st number	5	6	7	8
2nd number	15	18	21	

10. Write a rule for the pattern above. ____ top by ____.

11. 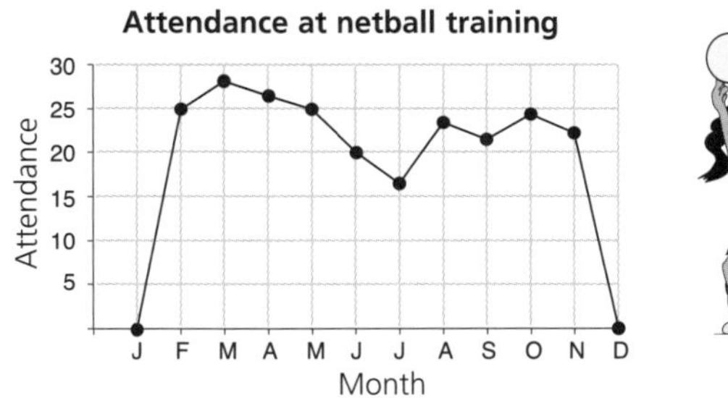

a The attendance in August. ____

b In which month did all 28 attend? ____

c Give a possible explanation for the January and December attendances.
____
____

27:4 Extension out of 6

1. a 4 × 159 = (4 × 100) + (4 × 50) + (4 × 9)
= ____

b 6 × 365 = (__ × __) + (__ × __) + (__ × __)
= ____

2. Days in 6 consecutive years. ____

3. 4, 11, 17, 25, 32, 39

a Which number is incorrect in the pattern above? ____

b If the pattern continued correctly, what would the 10th number be? ____

4. a 73 = ____ × ____ + ____

b 61 = ____ × ____ + ____

c 89 = ____ × ____ + ____

5. Use doubling to find:

a 2 × 536 ____ b 4 × 536 ____

6. If the shaded part has the value given, find the value of the whole.

a 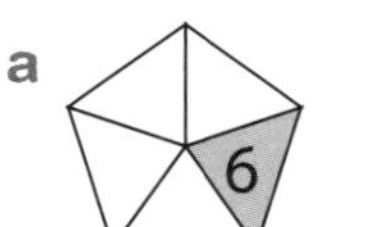

____

b 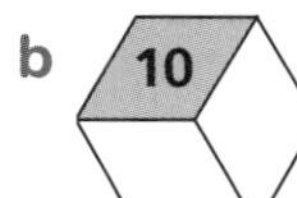

____

Challenge

Write facts about the number 45.

____
____
____
____
____

Concept

In each case finish the given pattern.

a Add 4.

23					

b Multiply by 2.

2					

c Subtract 9.

82					

d Add 11.

88					

e Multiply by 2 and add 1.

3					

f × 3 and then − 30.

16					

 • *AUSTRALIAN SIGNPOST MATHS NSW 5 MENTALS* • ISBN 978 0 6557 0912 1

28:1 out of 21

1. 60 × 5 ______
2. 0·1 × 10 ______
3. 142 − 19 ______
4. 352 − 19 ______
5. $\begin{array}{r} 6003 \\ +\ 2879 \\ \hline \end{array}$
6. 6 × ______ = 36
7. 8 × ______ = 32
8. 4·89, 4·9, 4·91, ______
9. 80 shared by 10. ______
10. $\begin{array}{r} \$50.08 \\ -\ \$41.97 \\ \hline \end{array}$
11. a 7 = ______ × ______ + ______
    b 11 = ______ × ______ + ______
    c 19 = ______ × ______ + ______
12. 3 × 46 = (3 × 40) + (3 × ______) = ______
13. Write the first 6 multiples of 5.
    ______
14. Three people donated $28. How much money was given altogether? ______
15. Round off 7 463 546 to the nearest:
    a million ______
    b thousand ______
16. Write the order from largest to smallest.
    3 359 574 3 395 637 3 392 035
    ______
17. Write 15:35 using 12-hour time. ______
18. Convert $\frac{18}{4}$ to a mixed number. $\square\frac{\square}{\square}$
19. 9 hours = ______ minutes
20. 5768 kilometres = ______ metres
21. 156, 166, 176, ______, ______, ______

28:2 out of 17

1. 0·7 × 100 ______
2. 586 − 246 ______
3. 954 − 299 ______
4. 435 + 285 ______
5. 8 × ______ = 56
6. 63 divided by 7. ______
7. 35 shared by 5. ______
8. 118 more than 354. ______

9.

	H	T	U
		8	1
×			3

10.

	H	T	U
	6	4	8
×			7

11.

	H	T	U
		3	8
×			3

12. 7 × 28 = (7 × 20) + (7 × 8) = ______
13.

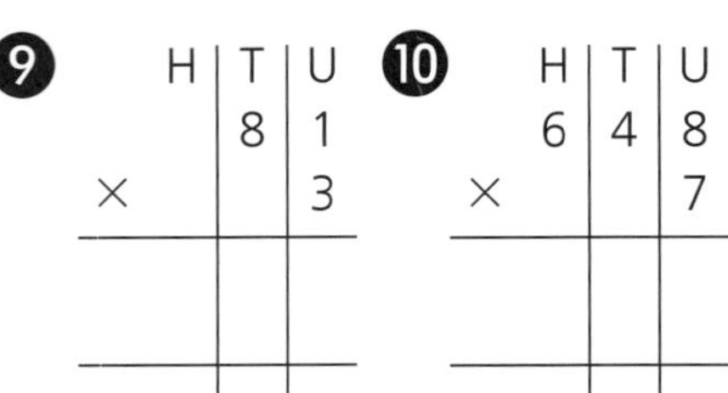

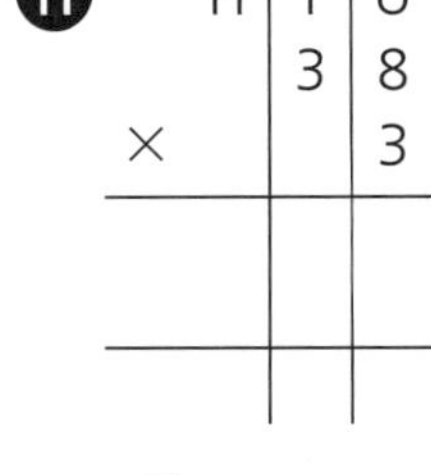

a When was my goldfish 3·5 cm long? ______
b How long was the fish on 1st July? ______

14. I bought 5 cartons of eggs with 18 eggs in each carton. How many eggs did I buy? ______
15. Write the first 6 multiples of 7.
    ______
16. How many groups of 4?

How many left over? ______ left over

17. 67 weeks = ______ days

For each row, circle the factors of the given number.

Number	Possible factors						
16	1	2	3	4	6	8	16
25	1	2	3	5	10	25	50
18	1	2	3	4	6	9	18
9	1	2	3	4	5	6	9

Number	Possible factors						
12	1	2	3	4	6	12	24
28	1	2	4	6	7	14	28
81	1	3	9	18	27	40	81
121	1	3	7	9	11	22	121

28:3 ☐ out of 13

1

H	T	U
3	6	4
×		2

2

H	T	U
	5	7
×		3

3

H	T	U
2	1	5
×		2

4 $3\overline{)993}$ **5** $2\overline{)862}$ **6** $5\overline{)655}$

7 Five people each baked 150 cakes for the cake stall. How many cakes were baked? ______

8 52 weeks = ______ days

9 9 × 35 = (9 × 30) + (9 × 5) = ______

10 These shapes are called

i______ shapes.

11

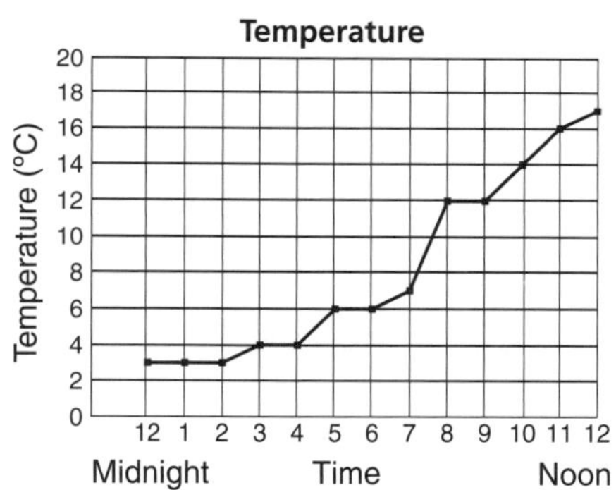

a What was the temperature at 5 am? ______

b At what time was it 14°C? ______

12 67·351 = ______ tens, ______ ones, ______ tenths, ______ hundredths and ______ thousandths

13 278 mm = ______ cm

28:4 ☐ out of 5 Extension

1 3, 12, 21, 30, ______

a What is the rule in the pattern above? ______

b What would the 9th number be? ______

2 Complete this table if the rule is *2nd number = 1st number* × 64 − 19.

1st number	1	2	3	4	5	6	7
2nd number							

3 **a** 4 × 426 = (4 × 400) + (4 × 20) + (4 × 6)

= ______ + ______ + ______

= ______

b 9 × 164 = (__ × __) + (__ × __) + (__ × __)

= ______ + ______ + ______

= ______

4 9 o'clock, quarter past 12, half past 9, 3 o'clock

Circle the above times where the hour and minute hands are at right angles.

5 200 minutes after 4:23 pm. ______

Challenge

Draw a diagram to show 329 multiplied by 7.

329 × 7 = ______

Can you balance this 25 kg mass using only the masses shown below?

Use the columns in the table to show as many answers as you can.

One answer is given.

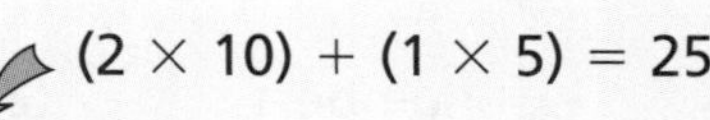

10 kg	2							
5 kg	1							
2 kg	0							
1 kg	0							

29:1 out of 19

1. 5 × 40 ____
2. 0·1 × 10 ____
3. $\frac{1}{2}$ of 48 ____
4. 100 − 26 ____
5. 145 + ____ = 200
6. 73 − ____ = 50
7. 8 × ____ = 32
8. 45 ÷ 5 ____
9. 48 × 8
10. 739 × 7
11. 528 × 9
12. a (3 × 6) + (2 × 6) ____
    b (4 × 5) + (3 × 7) ____
13. True or false, 6 kg < 4967 g. ____
14. 5 × 67 = (5 × 60) + (5 × ____) = ____
15. Which graph, A or B below, would show the growth of a seedling? ____

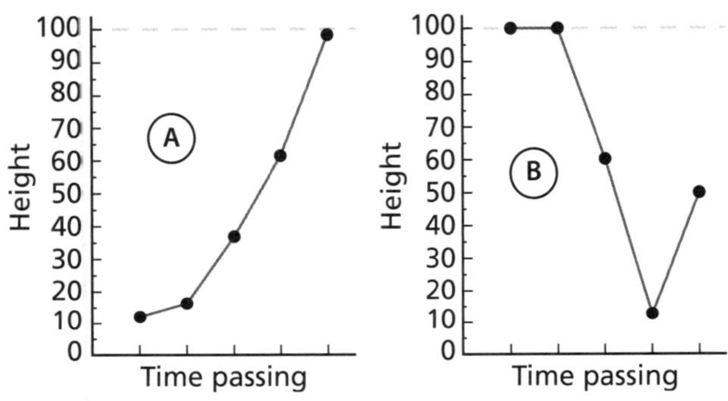

16. I drank a 250 mL can of kombucha four times this week. How much kombucha did I drink? ____
17. Round 5678 to the nearest:
    a thousand ____ b hundred ____
18. The average of 15 and 21 is ____.
19. 45 mm = ____ cm

29:2 out of 21

1. 40 ÷ 8 ____
2. 5 × 9 ____
3. 70 ÷ 10 ____
4. 4 × 7 ____
5. 300 × 8 ____
6. 0·4 × 100 ____
7. 8 squared. ____
8. 254 more than 187. ____
9. 37 × 5
10. 849 × 3
11. $7.66 × 8
12. How many threes can be taken from:
    a 25? ____ r ____ b 31? ____ r ____
13. Write $2\frac{31}{100}$ as a decimal. ____
14. 8 × 32 = (8 × 30) + (8 × ____) = ____
15. How much would each receive if we shared 1 kg of rice between:
    a 4 families? ____ b 5 families? ____
16. Round 4 567 935 to the nearest:
    a thousand ____
    b hundred ____
17. What is half of a half? ____
18. Rhiannon scored 7 runs, Tiegan 5, Stuart 9 and Pauline 8. What was their average number of runs? ____
19. Luke turns 56 in 13 years time. How old is he now? ____
20. I poured 375 mL water from a jug that held 2 L. How much water is left in the jug? ____
21. 9 years = ____ months

Turn to ID card A on page 6.

(20) 5 ____ 7
(21) 4 ____ 10
(22) 20 ____ 5
(23) 42 ____ 6
(24) ____ numbers
(25) ____ numbers
(26) ____
(27) 1, 8, ____ and ____
(28) ____
(29) ____

29:3 ☐ out of 12

1. $\begin{array}{r} \$1.06 \\ \times \quad 6 \\ \hline \end{array}$

2. $\begin{array}{r} 6849 \\ \times \quad 3 \\ \hline \end{array}$

3. $\begin{array}{r} 1676 \\ \times \quad 8 \\ \hline \end{array}$

4. $10\overline{)200}$

5. $10\overline{)340}$

6. $10\overline{)703}$ r

7. 276·93 has ______ hundreds, ______ tens, ______ ones, ______ tenths and ______ hundredths

8. 7 × 84 = (7 × 80) + (7 × ____) = ______

9. **People at the pool**

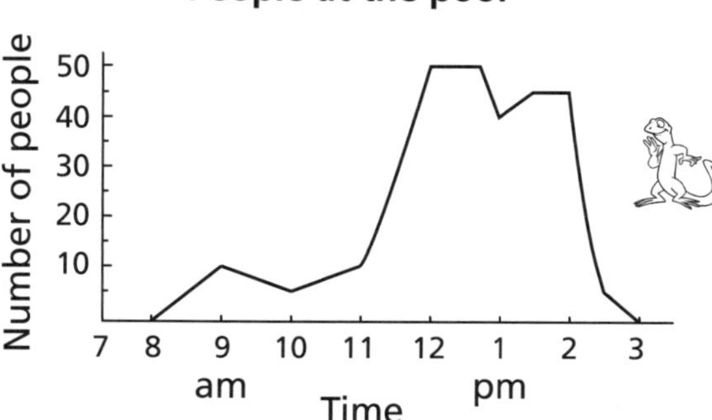

a How many people at 9 am? ______

b What was the maximum number of people at the pool? ______

During which hour did the number:

c increase the most? ______

d decrease the most? ______

10. I need 1200 g of tomatoes for my recipe. How many cans of 400 g tomatoes would I need? ______

11. Round 8 976 567 to the nearest:

a million ______ b ten ______

29:4 Extension ☐ out of 6

1. Akash saved \$376 a day. How long would it take him to save \$1600? ______

2. How many months in 14 years? ______

3. What fraction is shaded?

a  ______

b 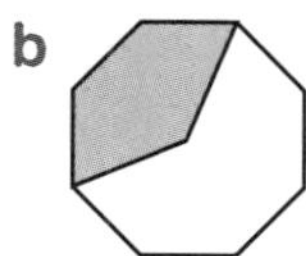 ______

4. The Roman numeral for 12. ______

5. I need 8 cubes to make a 2 × 2 × 2 stack like this. How many cubes do I need to make a 3 × 3 × 3 stack? ______

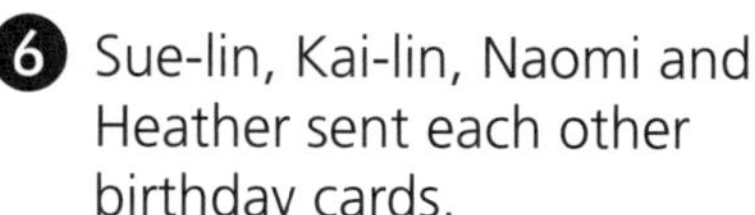

6. Sue-lin, Kai-lin, Naomi and Heather sent each other birthday cards.

How many cards were sent? ______

Challenge

Write facts about the number 603 730.

______

______

______

______

______

______

Roman numerals

1	I	one finger	
5	V	one hand	
10	X	two Vs	
50	L	half of a C	
100	C	centum = 100	
500	D	half of an ⓘ	
1000	M	mille = 1000	

IX = 10 − 1 = 9 | XC = 100 − 10 | CD = 500 − 100

1 Write our numeral for:

a XXXIX ______ b XL ______

c CCCIV ______ d CD ______

e DCXXI ______ f MDLI ______

2 Write the Roman numeral for:

a 65 ______ b 240 ______

c 966 ______ d 2500 ______

e 392 ______ f 1999 ______

30:1 ☐ out of 20

1. 8×5 ______
2. $560 + 37$ ______
3. $256 - 50$ ______
4. $\frac{1}{2}$ of 26. ______
5. $\begin{array}{r} 3685 \\ +\,3678 \\ \hline \end{array}$
6. $100 \times 0{\cdot}1$ ______
7. 4 multiplied by 20. ______
8. 88 divided by 8. ______
9. Share 12 between 4. ______
10. $\begin{array}{r} 9453 \\ -\,3671 \\ \hline \end{array}$
11. Estimate 33 + 59.
    E = ______ + ______
    E = ______
12. Estimate 164 + 38.
    E = ______ + ______
    E = ______
13. I carried 5 cans, each with a mass of 250 g. What was the total mass? ______
14. **a** $(10 \times 10) + (10 \times 4)$ ______
    b $(7 \times 10) + (7 \times 4)$ ______
15. Why is $\frac{1}{8}$ less than $\frac{1}{4}$?
    ____
    ____
16. Round 647 to the nearest hundred. ______
17. Morgan owes his father $90. If he pays $10 a week, for how many weeks will he have to pay? ______
18. What is the cost of 4 balls for $12.50 each? ______
19. 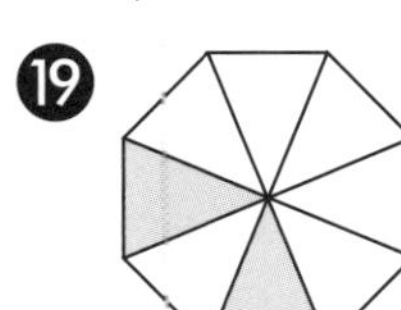
    What is the chance, as a fraction, that the spinner will land on grey? ______
20. Write 36 mm as centimetres. ______

30:2 ☐ out of 20

1. $265 - 248$ ______
2. $365 - 347$ ______
3. $0{\cdot}4 \times 10$ ______
4. 5×70 ______
5. $\begin{array}{r} 935 \\ \times \quad 5 \\ \hline \end{array}$
6. 81 divided by 9. ______
7. $6 \times$ ______ $= 54$
8. 45 divided by 5. ______
9. 354 more than 97. ______
10. $\begin{array}{r} 3579 \\ \times \quad 4 \\ \hline \end{array}$
11. Estimate 7354 + 98.
    E = ______ + ______
    E = ______
12. Estimate 3275 − 838.
    E = ______ − ______
    E = ______
13. How many 450 mL cups of water could I pour from a jug holding 2 L? ______
14. Write the Roman numeral for 8. ______
15. 300 mL of milk is left in a 2 L milk container. How much has been used? ______

16. How many 250 mL glasses can I fill with 3 L of water? ______
    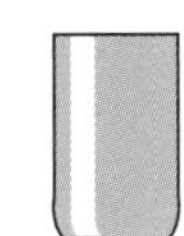
17. $456 \times 4 \times 0 \times 26$ ______
18. Round 850 to the nearest hundred. ______
19. If '<' means 'is less than' and '>' means 'is greater than', choose the sign that makes the number sentence true.
    a $6\frac{1}{2} \triangle 7$? $\triangle =$ ______
    b $0{\cdot}20 \triangle 0{\cdot}9$? $\triangle =$ ______
20. $6{\cdot}354$ kg = ______ grams

Averages

Measure the pictures in mm to obtain your answers.

a

Average length = ______
Average height = ______

b

Average mass = ______
Average width = ______

c

2L
250 mL
1L
550 mL

Average capacity = ______

Average height = ______

© PEARSON AUSTRALIA 2024 • *AUSTRALIAN SIGNPOST MATHS NSW 5 MENTALS* • ISBN 978 0 6557 0912 1

30:3 ☐ out of 14

1

H	T	U
	8	3
×		6

2

H	T	U
3	7	1
×		7

3

H	T	U
	8	4
×		6

4 Estimate 374 − 49.
E = ______ − ______
E = ______

5 Estimate 846 + 357.
E = ______ + ______
E = ______

6 A container of olives cost $3.80.
How much would 4 containers cost? ______

7 Write the Roman numeral for 4. ______

8 1339 + 43 ______

1339

9 I scored 68% in a test with 100 questions.
How many questions did I get wrong? ______

10 Is 34 × 6 closer to 30 × 6, 40 × 6 or 30 × 10?

______

11 Estimate the cost of 4 toys, each costing $34.56. ______

12 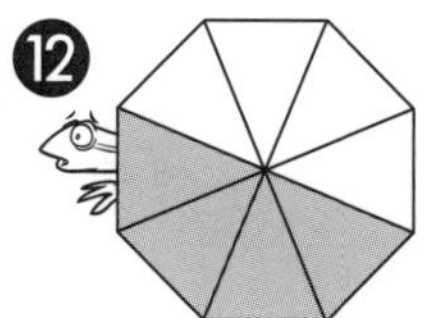
What is the chance, as a fraction, that the spinner will land on grey? ______

13 150 cm − 80 cm ______

14 Use the short form to write:

a two hundred grams ______

b 47 millilitres ______

30:4 Extension ☐ out of 5

1 Amelia bought 3 packets of cucumbers for $4.25 each and Sarah bought 4 packets of carrots for $2.70 each.
What was the total cost? ______

2 I make rings that have a mass of 25 g, 58 g, 73 g or 99 g.
I weighed four of the rings.
Their total mass was 318 g.
Which rings were weighed?

______

3 What is the chance, as a fraction, that the spinner will land on:

a A? ☐ **b** B? ☐ **c** C? ☐

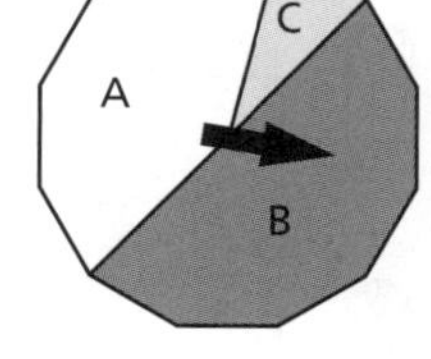

4 An astronaut can orbit a planet four times in a month. How many times could he orbit it in 8 years?

______

5 How many diagonals on 25 hexagons?

______

Challenge

Write your own number patterns and the rule for each.

______
______
______

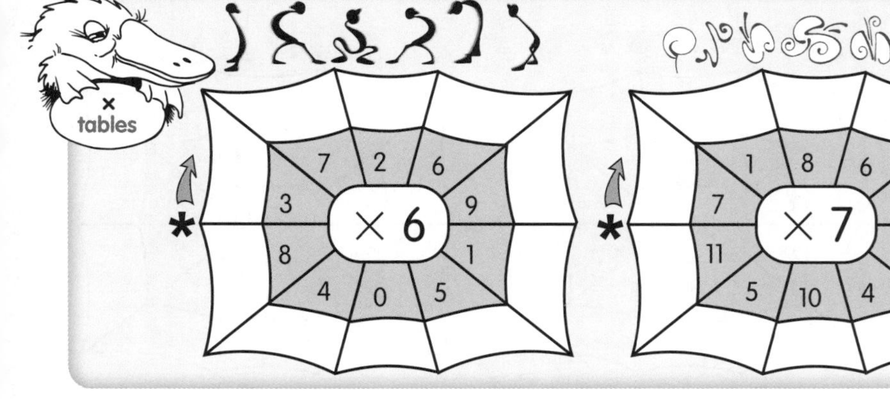

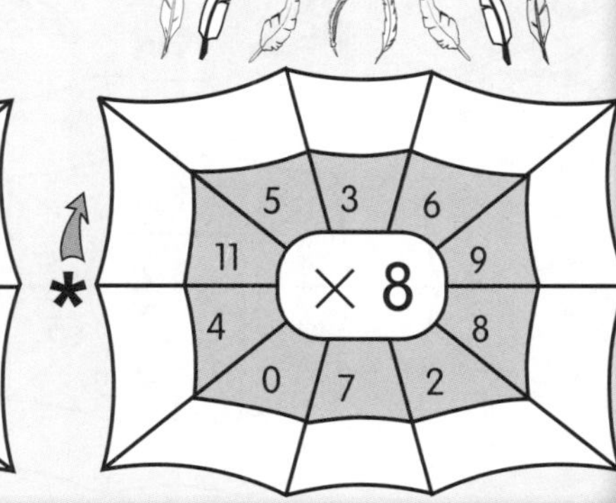

31:1 — out of 19

1. 45 + ______ = 100
2. 89 + ______ = 100
3. 6 × 4 ______
4. 58 − 24 ______
5.
```
  3856
+ 2645
```
6. 4 times 5. ______
7. 24 divided by 6. ______
8. \$57 − \$23 ______
9. 56 m + 23 m ______
10.
```
  $56.39
− $28.53
```

11. a 2 hours = ______ minutes
    b 3 years = ______ months
12. I scored 89 out of 100 in a test. How many more marks did I need to score 100? ______
13. How many nails in three packets of 32? ______
14. 4, 8, 12, ______, ______, ______, ______
15. Metres in 6 kilometres. ______
16. What is the chance, as a fraction, that the spinner will land on:
    a white? ______
    b grey? ______
17. Use am or pm to write 13:43. ______
18. Sam has 20 books and Gino has 3 times as many. How many books do the boys have altogether?

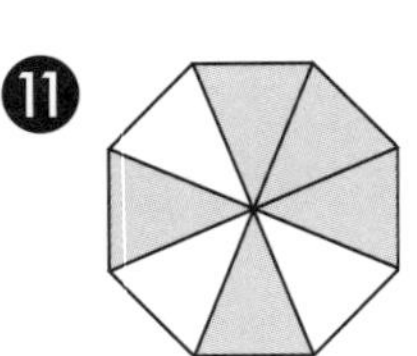

______

19. An exercise book costs \$2.40. Circle the best estimate for the cost of 9 books.

\$10 \$20 \$30 \$50

31:2 — out of 18

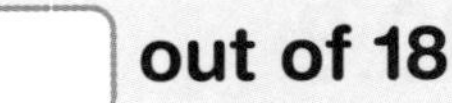

1. 28 ÷ 7 ______
2. 67 + ______ = 100
3. 5 × ______ = 40
4. 30 ÷ 6 ______
5.
```
  5746
×    7
```
6. 4 squared. ______
7. 36 divided by 6. ______
8. 5 times 600. ______
9. 72 shared by 9. ______
10.
```
  $39.63
×      4
```
11. What is the chance, as a fraction, that the spinner will land on:
    a white? ______
    b grey? ______
12. Write 3 tenths as a decimal. ______
13. Write 78 hundredths as a fraction. $\frac{\square}{\square}$
14. 7564 mL = ______ L
15. a What fraction of these results are odd? ______

Even	𝍸 𝍸 𝍸 IIII
Odd	𝍸 𝍸 𝍸 I

    b On a die, is the chance of throwing an odd or even number the same? ______
16. a How many pairs of birds, if I have 12 males and 12 females? ______

    b How many pairs of gloves, if I have 18 gloves? ______
17. Write the fraction equal to zero point one. $\frac{\square}{\square}$
18. 54, 48, 42, ______, ______, ______, ______

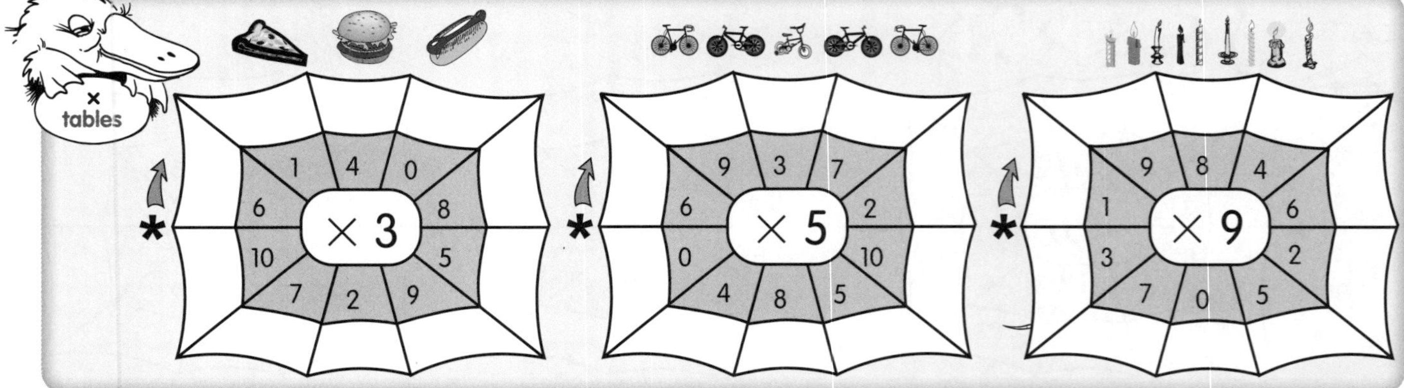

31:3 ☐ out of 9

1. 536 × 3

2. 7436 × 7

3. 15 kg 15 kg 15 kg 75 kg

   How many small Vikings are needed to balance the large one? ______

4. What is the probability, as a fraction, of tossing a 3 on a regular dice? $\frac{\square}{\square}$

5. Write 33 hundredths as a decimal. ______

6. How many times can a ribbon 11 cm long be cut from a length of 60 cm? ______

7. 5 red and 4 green marbles are in a bowl.
   What is the chance, as a fraction that a marble chosen would be:
   a red? ______
   b green? ______

8. I tossed a coin many times. Here are my results.

   a How many times did I toss a coin? ______

	Throwing a coin	Total
heads	卌 卌 卌 ||	
tails	卌 卌 卌 |||	

   b What fraction of the time did I toss a:
   head? $\frac{\square}{\square}$ tail? $\frac{\square}{\square}$

9. Write 78 million. ______

31:4 ☐ out of 5

Extension

1. The length of straw can be 7·8 cm, 8·8 cm or 9·8 cm. What two straws have a total length of 16·6 cm? ______

2. An octagon with the head of a tiger inside it, represents 80 tigers.
   What number is represented below:
   ______

3. I bought at least one of each of these items. The total cost was $3.25. How many oranges did I buy?
   $1.10
   35c
   ______

4. A truck can carry 10 tonnes. How many trips would be needed to carry 125 tonnes?
   ______

5. What is the chance, as a fraction, that the spinner will land on:
   A B C D
   a A? ______ **b B**? ______

Challenge

List events that have a chance of $\frac{1}{2}$.

______
______
______

Concept

Codes

Use this code to write *true* or *false* for each code statement.

A	B	C	D	E	F	G	H	I	J	K	L	M	N	O	P	Q	R	S	T	U	V	
4	9	6	?	/	2	$	+	5	<	&	0	>	7	#	1	=	8	3	*	(	)	•

*8(/ means 'true'.

a *+8//•>#8/•*+47•757/•53•/=(40•*#•*+58*//7. ______

b 641*457 •6##& •64>/•28#> •/7$047? •0#7$ •4$#. ______

c 400 •6#(7*57$ •7(>9/83 • +4)/ •#7/ •43 •4 •246*#8. ______

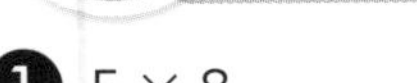

32:1 out of 19

1. 5×8 ____
2. $45 \div 9$ ____
3. $135 + 15$ ____
4. $200 - 45$ ____
5. $80 +$ ____ $= 100$
6. 0·3, 0·4, 0·5, ____
7. 24 divided by 6. ____
8. 70 shared by 7. ____
9. 5)405
10. 4)$784
11. 2)482
12.  
   a What fraction is shaded? 
   b What fraction is not shaded? 
13. True or false? $\frac{3}{10} = \frac{2}{5}$ ____
14. Would you use millilitres or litres to measure the water in a bathtub? 
   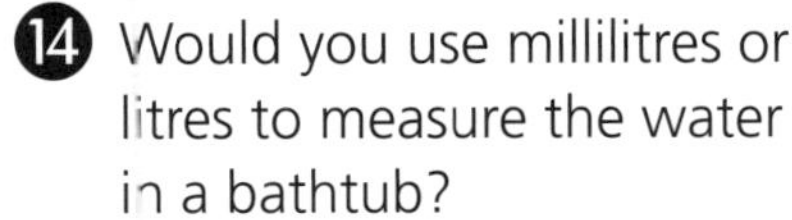
15. Order these fractions from largest to smallest.
   $\frac{1}{8}$ $\frac{1}{3}$ $\frac{3}{4}$ $\frac{8}{9}$ 
16. a $1 - \frac{1}{6}$ ____ 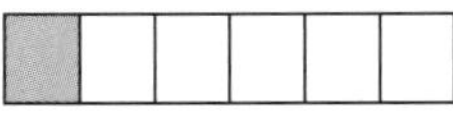
   b $1 - \frac{1}{4}$ ____ 
17. How many litres are in 6000 mL? ____
18. Share 27 pens among 3 girls.
   One share = ____
19. Months in 5 years. ____ 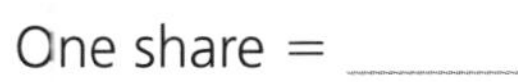

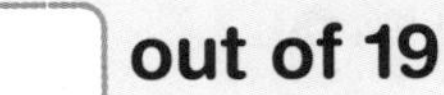

32:2 out of 19

1. 9×7 ____
2. $42 \div 6$ ____
3. $197 - 56$ ____
4. $435 + 185$ ____
5. 1·4, 1·5, 1·6, ____
6. 48 divided by 8. ____
7. $7 \times$ ____ $= 49$
8. $850 - 250$ ____
9. 2)683
10. 6)$594
11. 8)805
12. Use the diagram to find:
   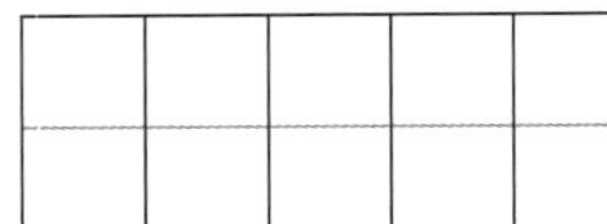
   a $1 - \frac{7}{10} = \frac{\square}{\square}$
   b $1 - \frac{3}{5} = \frac{\square}{\square}$
13. True or false? $\frac{8}{10} = \frac{4}{5}$ ____
14. Write the 24-hour time for 6:59 am. ____
15. Order these fractions from smallest to largest.
   $1\frac{1}{2}$, $\frac{1}{2}$, $1\frac{3}{10}$, $\frac{1}{5}$ ____
16. 15 mL  60 mL 
   How many caps of water are needed to fill the eggcup? ____
17. Write 9375 mL as litres and millilitres. ____
18. Make the denominators equal, then add.
   $\frac{1}{4} + \frac{5}{8} = \frac{\square}{\square} + \frac{\square}{\square} = \frac{\square}{\square}$
19. What is the abbreviation for:
   a litre ____ b millimetre ____
   c metre ____ d square metre ____

Turn to ID card B on page 7.
Give the answers for these numbers.

(6) ____ of an angle (7) ____ of an angle
(8) ____ angle (9) ____ angle
(10) ____ angle (11) ____ angle
(12) ____ angle (13) ____
(18) axis of ____ (19) ____ of symmetry

 ISBN 978 0 6557 0912 1

32:3 ☐ out of 15

1 $7\overline{)835}$

2 $2\overline{)306}$

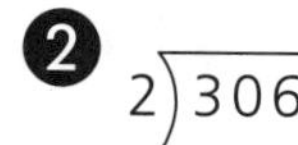

3 $9\overline{)693}$

4 $\begin{array}{r} 2467 \\ \times \quad 6 \\ \hline \end{array}$

5 $\begin{array}{r} 8436 \\ \times \quad 8 \\ \hline \end{array}$

6 **a** $1 - \frac{2}{3}$ ______ **b** $1 - \frac{6}{10}$ ______

7 Order these fractions from smallest to largest.

$\frac{3}{4}$ $\frac{7}{4}$ $\frac{1}{2}$ $\frac{1}{4}$

8 True or false? $\frac{10}{16} = \frac{6}{8}$ ______

9 Make the denominators equal, then subtract.

$\frac{4}{5} - \frac{1}{10} = \square - \square = \square$

10 What is the total quantity held by these bottles? ______

11 How many millilitres are in 5·9 L? ______

12 Write 7359 as litres and millilitres. ______

13 Name a container that holds about 1 litre.

______

14 Circle the larger decimal in each part.

a 0·27 or 0·4 **b** 0·71 or 0·7

15 Write 0·52 as a:

a percentage ______ **b** fraction

32:4 Extension ☐ out of 7

1 **a** 2·5 L = ______ mL **b** 800 mL = ______ L

2 If the length, width and height of this model are all multiplied by 2, how many cubes will be used in the new model's:

a top view? ______

b total volume? ______

3 0·16, 0·18, 0·2, ______, ______, ______

4 The Roman numeral for:

a 26 ______ **b** 52 ______

5 **a** 147 + ______ = 264

b 674 − ______ = 283

6 In every space made by a row of 6 trees, we planted 8 flowers.

How many flowers did we plant? ______

7 **a** 367 + 396 = ______ + ______ = ______

b 847 + 448 = ______ + ______ = ______

c 925 − 308 = ______ − ______ = ______

Write what you know about the number 3 564 009.

______

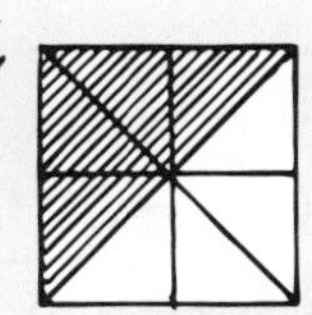

There are two eighths in 1 quarter.

Use the diagram to answer these questions.

a How many eighths in one half? ______

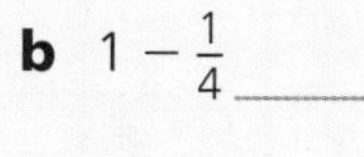

b $1 - \frac{1}{4}$ ______ **c** $\frac{1}{2} + \frac{1}{8}$ ______ **d** $\frac{5}{8} - \frac{1}{8}$ ______

e $\frac{3}{4} - \frac{3}{8}$ ______ **f** $\frac{1}{2} + \frac{5}{8}$ ______ **g** $\frac{7}{8} - \frac{3}{4}$ ______

h $1\frac{7}{8} - \frac{1}{4}$ ______ **i** $2 - \frac{1}{4}$ ______ **j** $1\frac{1}{2} - \frac{5}{8}$ ______

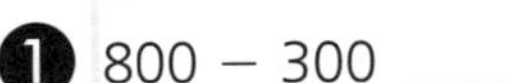

33:1 ☐ out of 20

1. 800 − 300 ______
2. 100 − 57 ______
3. 46 + 19 ______
4. 60 + ______ = 92
5. $\begin{array}{r} 7465 \\ +\ 1756 \\ \hline \end{array}$
6. 6 × ______ = 36
7. 9 × ______ = 27
8. 12 ÷ 4 ______
9. 4 groups of 20. ______
10. $\begin{array}{r} 1546 \\ \times\quad 6 \\ \hline \end{array}$
11. **a** $\frac{3}{4}$ plus $\frac{1}{4}$. ☐☐☐☐ ______

    b 1 minus $\frac{1}{4}$. ______
12. How many millilitres are in 5 L? ______
13. Order these fractions from smallest to largest.

    $\frac{5}{8}$ $\frac{1}{2}$ $\frac{3}{8}$ $\frac{1}{4}$

14. How many millimetres in 8 cm? ______
15. If 32·5 million is 32 500 000, what is:

    26·5 million? ______
16. How many $4 balls can I buy for $16? ______
17. Would you use millilitres (mL) or litres (L) to measure the capacity of a teaspoon? ______
18. I drank three quarters of my drink. What fraction of my drink is left? $\frac{\square}{\square}$
19. **a** 8300 mm = ______ m **b** 0·25 = ______ %

    c 1845 g = ______ kg **d** 0·56 = ______ %
20. Make the denominators equal then add.

    $\frac{3}{4} + \frac{2}{8} = \frac{\square}{\square} + \frac{\square}{\square} = \frac{\square}{\square}$

33:2 ☐ out of 18

1. 49 + 257 ______
2. 857 − 359 ______
3. 78 + ______ = 200
4. 56 ÷ 7 ______
5. $\begin{array}{r} 8465 \\ -\ 3646 \\ \hline \end{array}$
6. 8 multiplied by 9. ______
7. 49 divided by 7. ______
8. 6 × ______ = 54
9. 593 − ______ = 492
10. $\begin{array}{r} 2546 \\ \times\quad 4 \\ \hline \end{array}$
11. **a** 7·5 m = ______ mm **b** 4836 g = ______ kg
12. Write the numeral 64·7 million. ______
13. **a** $1 - \frac{4}{5}$ $\frac{\square}{\square}$ **b** $\frac{4}{5} - \frac{2}{5}$ $\frac{\square}{\square}$
14. Make the denominators equal then add.

    $\frac{5}{6} - \frac{5}{12} = \frac{\square}{\square} - \frac{\square}{\square} = \frac{\square}{\square}$
15. Millilitres in $5\frac{1}{2}$ L. ______
16. Find the total capacity. ______

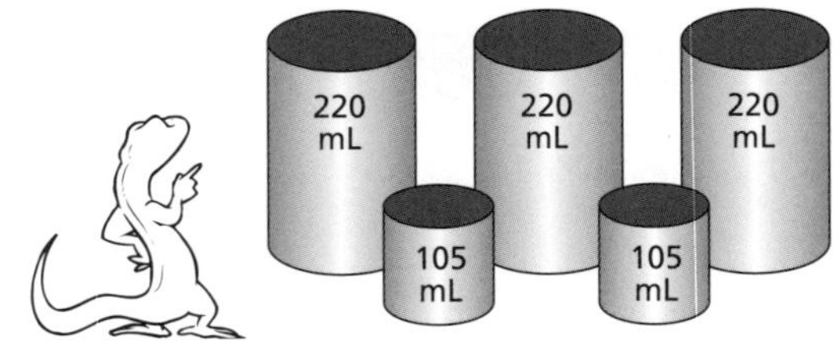

17. Jessica had 1 metre of ribbon. She cut two ribbons off, each 30 cm.

    a How much ribbon was left? ______

    b What fraction of the ribbon was left? $\frac{\square}{\square}$

    c What percentage of the ribbon was left? ______
18. What change do I get from $50 if buy 3 bears? ______

Concept

In a magic square, the sum of each row, column and diagonal is the same.

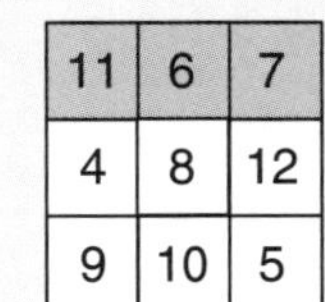

11	6	7
4	8	12
9	10	5

Here, all lines add up to 24.

Complete these magic squares.

a

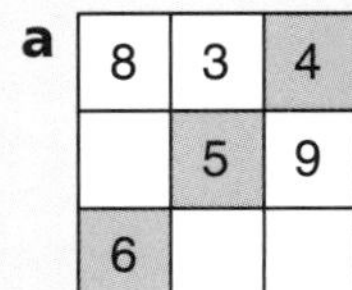

8	3	4
	5	9
6		

b

9		5
2		10
7		3

c

8		
9	7	
		6

 • *AUSTRALIAN SIGNPOST MATHS NSW 5 MENTALS* • ISBN 978 0 6557 0912 1

33:3 ☐ out of 15

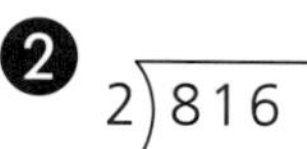
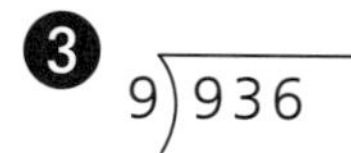

1. $5\overline{)547}$ 2. $2\overline{)816}$ 3. $9\overline{)936}$

4. $\begin{array}{r} 3697 \\ +2467 \\ \hline \end{array}$ 5. $\begin{array}{r} 4367 \\ \times\quad 8 \\ \hline \end{array}$

6. a $5 - \frac{3}{10} = \frac{\square}{\square}$ b $8 - \frac{4}{10} = \frac{\square}{\square}$

7. True or false? $\frac{4}{12} = \frac{2}{6}$ ______

8. Write 5 L 354 mL:
   a in millilitres. ______
   b as litres, using a decimal. ______

9. 26 weeks = ______ days

10. Order these fractions from smallest to largest.
    $\frac{1}{8}, 1\frac{4}{5}, 1\frac{1}{10}, \frac{1}{4}$ ______

11.  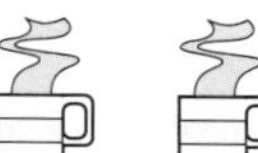 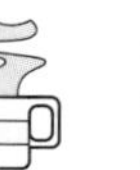 
    0 mL   10 mL   80 mL   140 mL   240 mL
    What is the average amount in each mug? ______

12. a 23·67 L = ______ mL   b 0·87 = ______ %
    c 3256 m = ______ km   d 0·13 = ______ %

13. Reuben owes his mum $150. If he pays back $10 each week, how long will it take him to repay her? ______

14. Michael earned $110 but spent $17.25 on fares and $38.25 on food. How much does he have left? ______

15. 8·9, 9·0, 9·1, ______, ______, ______, ______

33:4 Extension ☐ out of 5

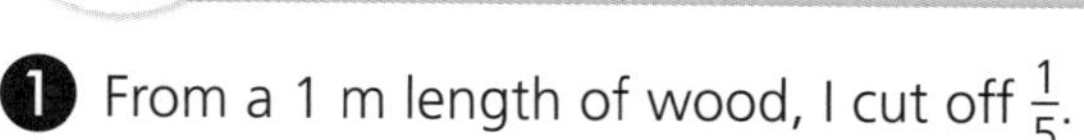

1. From a 1 m length of wood, I cut off $\frac{1}{5}$.
   a What fraction of the wood is left? $\frac{\square}{\square}$
   b What length, in centimetres, is left? ______

2. The Roman numeral for:
   a 32 ______   b 60 ______

3. If 18·8 billion is 18 800 000 000, what is:
   a 73·2 billion? ______
   b 69·1 billion? ______

4. The masses of four boxes are 3 kg, 580 g, 1 kg 56 g and 2300 g.

   What is the total mass? ______

5. a I can travel 7 km on a litre of fuel. How much petrol would I use to travel 91 km? ______
   b What would the cost be if I paid 210 cents per litre? ______

Challenge

Complete this fraction wall. Record equivalent fractions below.

$\frac{1}{2}$							
$\frac{1}{4}$							
$\frac{1}{8}$							

Note: $\frac{2}{4} = \frac{4}{8}$

Use the diagram to answer these questions.

a $1 + \frac{1}{4} =$ ______   b $\frac{1}{2} + \frac{1}{4} =$ ______   c $\frac{1}{2} - \frac{1}{4} =$ ______

d $1 - \frac{1}{4} =$ ______   e $\frac{3}{4} - \frac{1}{4} =$ ______   f $\frac{3}{4} - \frac{1}{2} =$ ______

g $\frac{2}{4} + \frac{1}{2} =$ ______   h $1 - \frac{3}{4} =$ ______   i $2 - \frac{1}{2} =$ ______

j $\frac{3}{4} + \frac{1}{2} =$ ______   k $\frac{3}{4} + \frac{3}{4} =$ ______   l $2 - \frac{3}{4} =$ ______

 • *AUSTRALIAN SIGNPOST MATHS NSW 5 MENTALS* • ISBN 978 0 6557 0912 1

34:1 out of 18

1. 320 + 85 ____
2. 400 − 90 ____
3. 7 × 6 ____
4. 5 × 8 ____
5. $\begin{array}{r} 6854 \\ +\,2973 \\ \hline \end{array}$
6. 8 × ____ = 24
7. 6 × ____ = 30
8. 20 ÷ 5 ____
9. 24 ÷ 8 ____
10. $\begin{array}{r} 3640 \\ \times\quad 7 \\ \hline \end{array}$
11. Bananas cost 70 cents each. How much do 5 bananas cost? ____
12. I sold 20 drinks in 5 minutes. On average, how many were sold in 1 minute? ____

13. 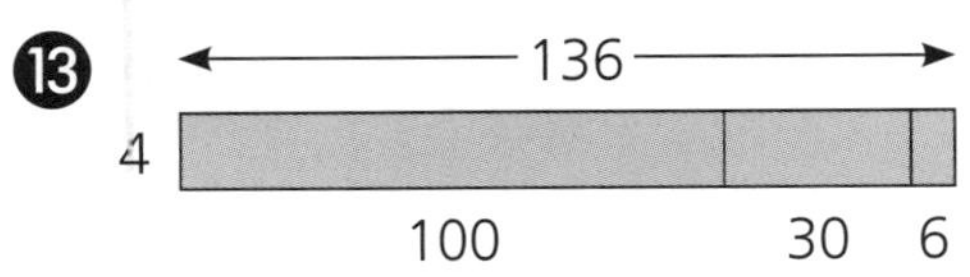

4 × 136 = (4 × ___) + (4 × ___) + (4 × ___)
= ____ + ____ + ____
= ____

14. **a** 0·68, 0·69, 0·70, ____, ____, ____
    b The rule for this pattern is ____.
15. 5·6 million as a numeral. ____
16. **a** 0·356 = ____ % **b** 7·1 L = ____ mL
17. 500 ha = ____ square kilometres
18. Tiffany had $40. She spent:
    - on food $14.50
    - at the movie $ 7.00
    - on stamps $ 2.25

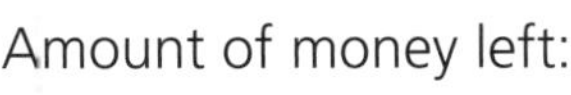

Tiffany's total: ____

Amount of money left: ____

34:2 out of 18

1. 8 × 7 ____
2. 9 × 6 ____
3. 54 ÷ 9 ____
4. 49 ÷ 7 ____
5. $\begin{array}{r} 3805 \\ -\,2797 \\ \hline \end{array}$
6. 6 × ____ = 48
7. 8 × ____ = 72
8. 7 groups of 9. ____
9. Multiply 8 by 5. ____
10. $\begin{array}{r} 3186 \\ \times\quad 8 \\ \hline \end{array}$
11. $\frac{3}{8}$, $\frac{4}{8}$, $\frac{5}{8}$, ▭, ▭, ▭, ▭, ▭
12. Sally completed a test with 100 questions. Her result was 78%. What fraction of the questions did she get wrong? ▭
13. Write 2·9 billion as a numeral.
    ____
14. 7400 mL = ____ L
15. Round 637 to the nearest hundred. ____
16. Jacob had $26. He bought these items.

$2.60

$3

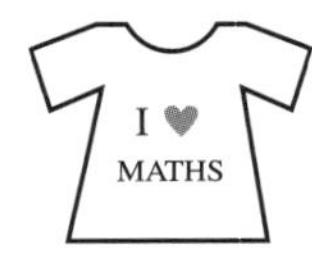

$15

Total amount spent: ____

Amount left: ____

17. 900 ha = ____ square kilometres
18. 5 × 527 = (__ × __) + (__ × __) + (__ × __)
    = ____ + ____ + ____
    = ____

427
5
400 20 7

Party time

Patrick's party started at 7 pm. In the first 5 minutes 3 friends came. In the next 5 minutes 8 friends came. In each 5 minute period until 7:30 pm 4 more friends came. How many friends had arrived by 7:30 pm?

____

 AUSTRALIAN SIGNPOST MATHS NSW 5 MENTALS • ISBN 978 0 6557 0912 1

34:3 ☐ out of 9

1. a $6\ km^2$ = ______ ha

   b $50\,000\ m^2$ = ______ ha

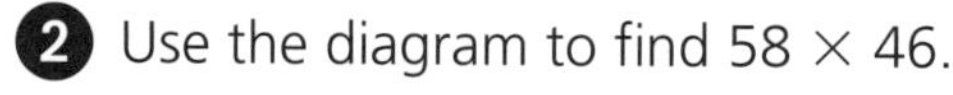

2. Use the diagram to find 58 × 46.

   = (__ × __) + (__ × __) + (__ × __) + (__ × __)
   = ______ + ______ + ______ + ______
   = ______

	50	8
40		
6		

3. A kite costs $12.30. How much will Megan have to pay for 3 kites?

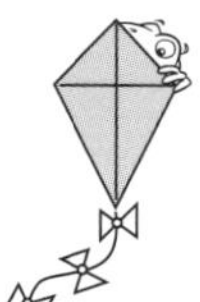

   ______

4. 130%, 110%, 90%, ______, ______, ______

   The rule for this pattern is ______________.

5. How many hectares in $60\,000\ m^2$? ______

6. Jason placed a rectangular prism on a table. One sixth of the faces were not visible. What fraction of the faces were visible? $\frac{\square}{\square}$

7. Write 7·2 million as a numeral.

   ____

8. a 8675 m = ______ km

   b 3400 mL = ______ L

9. Find an estimate by rounding each number first.

   a 58 × 34 ______ b 48 × 53 ______

   c 82 × 49 ______ d 27 × 38 ______

34:4 Extension ☐ out of 6

1. a 7 ha = ______ m^2

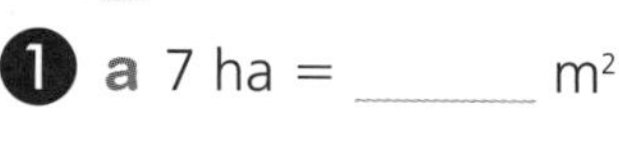

   b $25\ km^2$ = ______ ha

   c 35 ha = ______ m^2

2. The Roman numeral for:

   a 8 ______ b 65 ______

3. a $3\frac{3}{5}$, $4\frac{4}{5}$, 6, $\frac{\square}{\square}$, $\frac{\square}{\square}$, $\frac{\square}{\square}$, $\frac{\square}{\square}$

   b The rule for this pattern is ______________.

4. Sarah had 2 hexagonal prisms. She coloured one of the faces of the first prism. What fraction of the faces did she colour? $\frac{\square}{\square}$

5. Use an area model as in 34:3, Question 2 to find:

   a 73 × 68 ______ b 537 × 8 ______

6. If each full-sized cow is given exactly 1 hectare of space, describe two possible dimensions in metres of a paddock with 6 cows.

   ____

   ____

Challenge

Make up your own number patterns and write the rule for each.

____

____

____

____

____

Which total is easiest to throw using two dice?

a Which total would you choose? ______

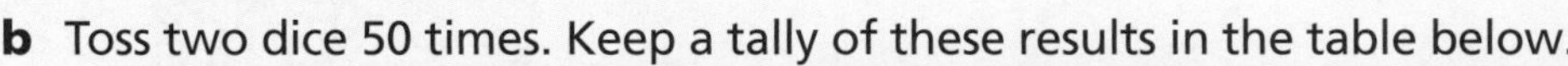

b Toss two dice 50 times. Keep a tally of these results in the table below.

Number	1	2	3	4	5	6	7	8	9	10	11	12
Tally												

c Which total occurred most often? ______

 • *AUSTRALIAN SIGNPOST MATHS NSW 5 MENTALS* • ISBN 978 0 6557 0912 1

35:1 out of 14

1. 574 + 99 ______
2. 376 − 99 ______
3. 7 × 6 ______
4. 5 × 7 ______
5. 6 times 8. ______
6. 7 squared. ______
7. 36 divided by 6. ______
8. 36 shared by 9. ______
9. 72 × 30 = (70 × ____) + (____ × ____)
   = ________ + ________
   = ________
10. Round 437 to the nearest hundred. ______
11. 1 hectare = 100 m × 100 m = 10 000 m²
   a 3 ha = ______ m²
   b 70 000 m² = ______ ha
12. 1 square kilometre = 1000 m × 1000 m
   = 1 000 000 m²
   a 6 km² = ________ m²
   b 900 ha = ________ km²
13. Would you use m² or ha to measure the area of a small farm? ______
14. Cars sold last week

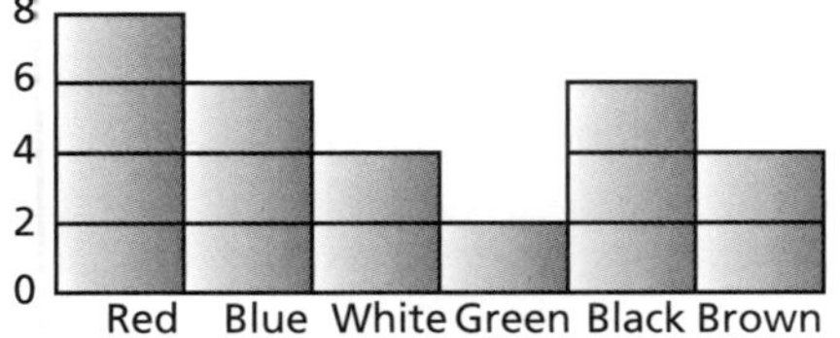

   a How many white cars? ______
   b What colour is least popular? ______
   c What colour is most popular? ______
   d How many blue, black and brown cars altogether? ______
   e How many cars altogether? ______

35:2 out of 14

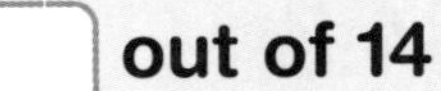

1. 645 − 297 ______
2. 257 + 397 ______
3. 81 ÷ 9 ______
4. 64 ÷ 8 ______
5. 8 squared. ______
6. 54 divided by 9. ______
7. Halve 364. ______
8. 0·4 × 100 ______
9. 24 × 20 = (20 × ____) + (____ × ____)
   = ________ + ________
   = ________
10. Use this diagram to find 58 × 46.

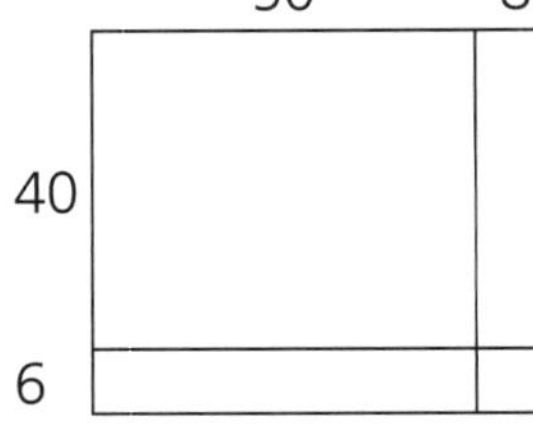

   = (__ × __) + (__ × __) + (__ × __) + (__ × __)
   = ______ + ______ + ______ + ______
   = ______
11.

   a What is the total cost of the star, the car and the drum? ______
   b How much change would you get from $10 if you buy the star and the car? ______
12. 5 hectare = ________ m²
13. 500 ha = ________ km²

14. Estimate by rounding each number first.
   a 53 × 89 ______ b 38 × 341 ______

Conduct a survey to collect information on a topic of interest.
For example, 'What is your favourite sport?'.

a Write a question for your chosen topic: ____________________

b Who will you ask? ____________________

c Record your results on separate paper. What did you find?

____

 ISBN 978 0 6557 0912 1

35:3 out of 10

1 13 × 64 = 64 × 10 ____ + 64 × 3 ____ = ____

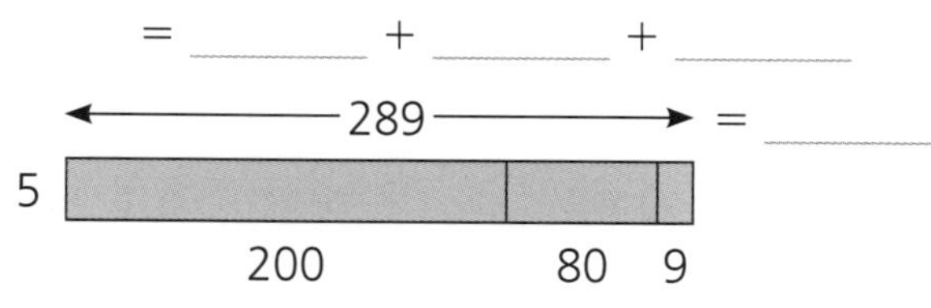

2 5 × 289 = (5 × ___) + (5 × ___) + (5 × ___)
= ____ + ____ + ____
= ____

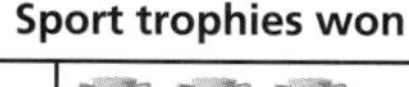

3 7·5 km = ____ m

4 Estimate by rounding each number first.
a 44 × 57 ____ b 26 × 552 ____

5 3 hectares = ____ m^2

6 a 4 km^2 = ____ m^2
b 800 ha = ____ km^2

7 Complete: 43 × 52 = (40 × ___) + (3 × ___)

8 Sport trophies won

a How many has Rajiv won? ____
b How many more has Mark won than Kari? ____
c How many have Rajiv and Kari won altogether? ____

9 I need 75g of raisins to make a batch of biscuits. What mass of raisins will I need to make 14 batches of biscuits? ____

10 Write $\frac{17}{10}$ as a mixed numeral. ____

35:4 out of 5

Extension

1 What is the least number of matchsticks needed to make:
a 2 hexagons? ____
b 7 hexagons? ____

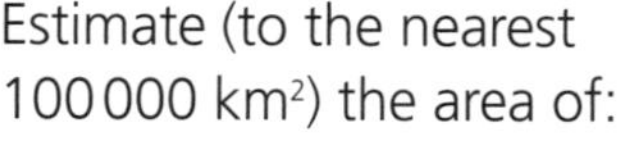

2 a Find an estimate for 68 × 47 by rounding off each number and then multiplying. E = ____
b Find the answer to 68 × 47. ____
c What is the difference between your estimate and the answer? ____

3 New South Wales has an area of 801 150 km^2.

Estimate (to the nearest 100 000 km^2) the area of:
a SA ____
b Qld ____

4 There were 50 people at the concert. Estimate the number of boys who attended. ____

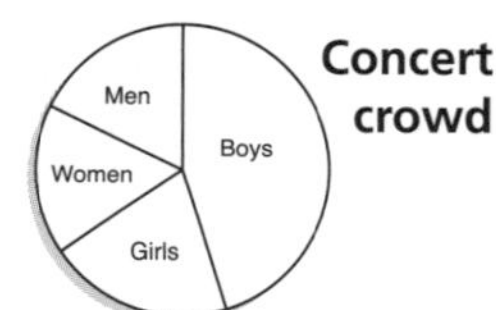

5 Use an area model, as in 34:3, Question 2 to find:
a 38 × 62 ____ b 385 × 5 ____

Challenge

Complete this fraction wall. Record equivalent fractions below.

$\frac{1}{2}$			$\frac{1}{2}$		
$\frac{1}{3}$					
$\frac{1}{6}$					

Note: $\frac{2}{3} = \frac{4}{6}$

Travel graph

a At how many *floors* did the lift stop? ____
b How long does the lift stop for at the 2nd floor? ____
c How many floors are in the building? ____
d How long did it take for the lift to return to the ground floor? ____

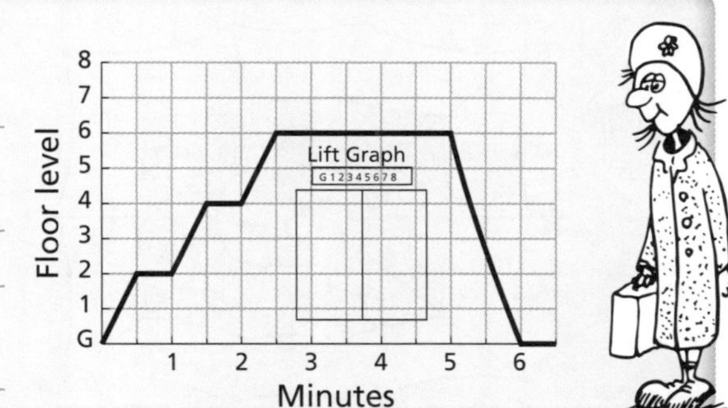

36:1 out of 15

1. $0{\cdot}1 \times 10$ ____
2. 4×70 ____
3. 6×30 ____
4. $175 - 25$ ____
5. $\begin{array}{r} 7564 \\ +\ 1546 \\ \hline \end{array}$
6. 50 divided by 5. ____
7. $4 \times$ ____ $= 36$
8. $5 \times$ ____ $= 45$
9. $42 \div 6$ ____
10. $\begin{array}{r} 6486 \\ +\ 3647 \\ \hline \end{array}$
11. \$10 minus \$3.50. ____
12. 6 hectare = ____ m²
13.

Tennis matches won

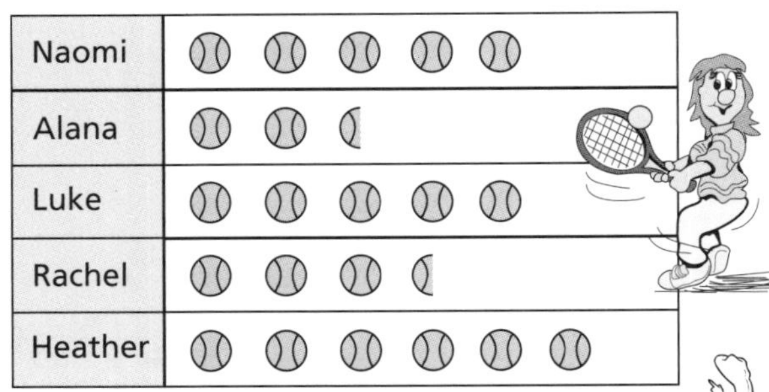

Naomi	5 balls
Alana	2½ balls
Luke	5 balls
Rachel	3½ balls
Heather	6 balls

(ball) = 4 matches

a How many matches were won by Alana? ____

b Which player won 14 matches? ____

c Which player won more than 20 matches? ____

14. Write 5 365 809 in words. ____

15. An obtuse angle is between 90° and 180°. What is a reflex angle between? ____

36:2 out of 15

1. $354 + 398$ ____
2. $784 - 357$ ____
3. $63 \div 7$ ____
4. $48 \div 8$ ____
5. 5·46, 5·47, 5·48, ____
6. $0{\cdot}7 \times 100$ ____
7. 42 divided by 7. ____
8. 56 shared by 7. ____
9. a $27 \times 91 = \begin{array}{r} 91 \\ \times\ 20 \\ \hline \end{array} + \begin{array}{r} 91 \\ \times\ 7 \\ \hline \end{array} =$ ____

   b $39 \times 17 = \begin{array}{r} 17 \\ \times\ 30 \\ \hline \end{array} + \begin{array}{r} 17 \\ \times\ 9 \\ \hline \end{array} =$ ____

10. Charlotte's change from \$5 if she bought an ice-cream for \$1.50? ____

11. How many faces has a triangular pyramid? ____

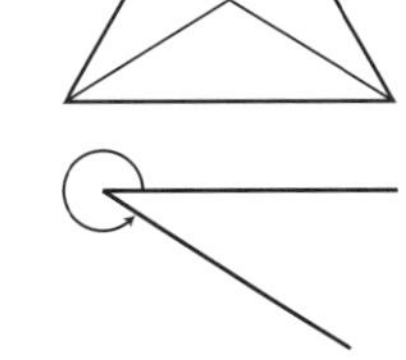

12. An angle larger than a straight angle is called a ____ angle.

13. If the acute angle above is 35°, what is the size of the reflex angle? ____
14. In a 100 m race, I fell after running 85 m.

    a How far was I from the end of the race? ____

    b What percentage of the race had I run? ____
15. 28 students in my class paid \$34 each for the excursion. How much was paid altogether? ____

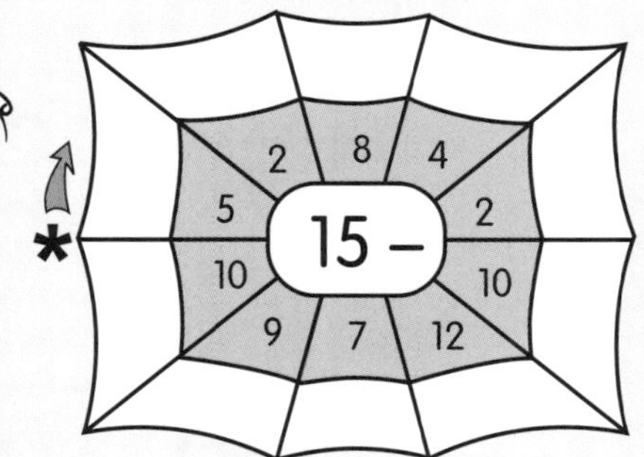

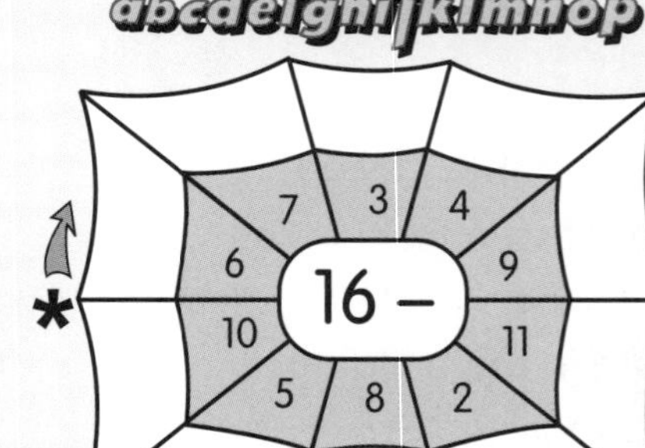

© PEARSON AUSTRALIA 2024 • *AUSTRALIAN SIGNPOST MATHS NSW 5 MENTALS* • ISBN 978 0 6557 0912 1

36:3 out of 13

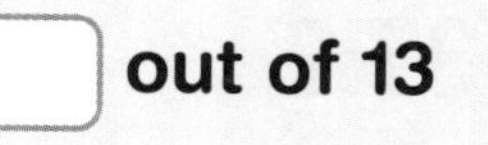

1. $\begin{array}{r} 23 \\ \times 11 \\ \hline \end{array}$

2. $\begin{array}{r} 43 \\ \times 21 \\ \hline \end{array}$

3. $\begin{array}{r} 65 \\ \times 43 \\ \hline \end{array}$

4. Estimate 27 × 12.
   E = ____ × ____
   = ____

5. Estimate 35 × 253.
   E = ____ × ____
   = ____

6. How many nails in 24 packets of 32? ____

7. What size is angle:
   a A? ____
   b B? ____

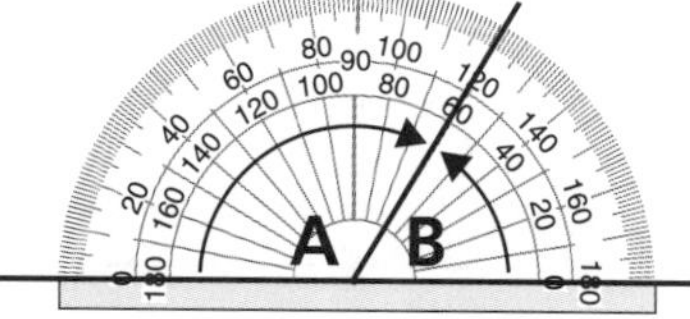

8. How many right angles are needed to make a revolution?

____

9. I have 3 pieces of string. Their lengths are 1 m 30 cm, 1 m 42 cm and 2 m 15 cm.
   What is the total length? ____

10. Grams in 3 kg 270 g. ____

11. Write the equivalent fraction for:
    a $\frac{1}{4}\frac{(\times 2)}{(\times 2)} = \frac{\square}{\square}$
    b $\frac{1}{3}\frac{(\times 4)}{(\times 4)} = \frac{\square}{\square}$

12. Write 9 204 665 in words. ____

13. Which part is the:
    a numerator? ____
    b denominator? ____

$$\frac{3}{4}$$

36:4 out of 6

Extension

1. Our class brought in fifty-seven 35 g glue sticks. Find the total mass. ____

2. What is the smallest number of matchsticks needed to make:

   a 2 small triangles? ____
   b 5 small triangles? ____

3. I spent $7.40 on four items like these. What did I buy?

____

4. This figure has squares of different sizes. How many squares altogether?

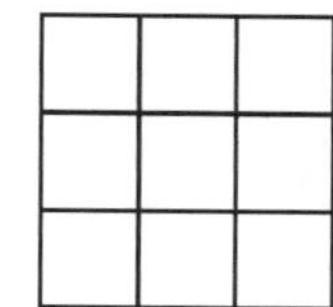

____

5. The Roman numeral for 95 is ____.

6. Our numeral for XCVIII is ____.

Challenge

Draw and label different types of angles.

 • *AUSTRALIAN SIGNPOST MATHS NSW 5 MENTALS* • ISBN 978 0 6557 0912 1

37:1 out of 16

1. 6 × 40 ______
2. 500 × 5 ______
3. 257 − 143 ______
4. 3 × 8 ______
5. 5 × ______ = 20
6. 6 × ______ = 36
7. 24 shared by 6. ______
8. 28 divided by 7. ______
9. $6.17 × 5
10. 8834 × 3
11. 3017 × 6

12. Which angle is:
 a reflex? ______
 b straight? ______
 c acute? ______
 d obtuse? ______

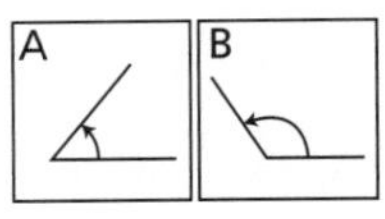

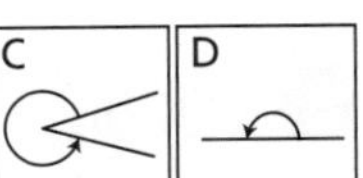

13. 
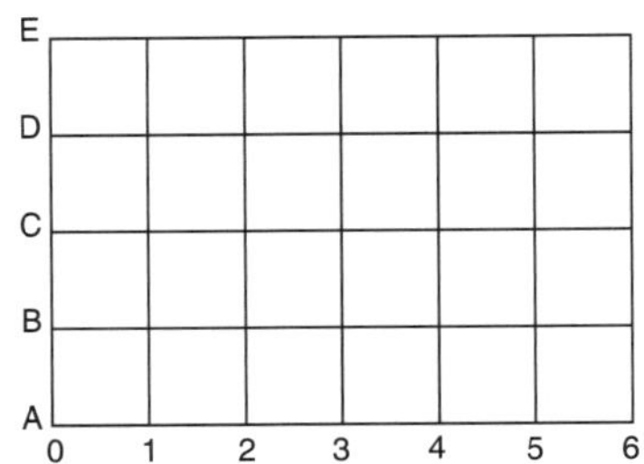

 a Plot this set of points on the grid and join them in order.
 1B, 2D, 3D, 4D, 5B, 4B, 3B, 2B, 1B
 b What shape is formed? ______

14. Write 903 413 in words.

______

______

15. What type of angle is between 0° and 90°? ______

16. Write the 24-hour time for 4:56 pm. ______

37:2 out of 17

1. 18 ÷ 6 ______
2. 28 ÷ 7 ______
3. 4 × 6 ______
4. 7 × 8 ______
5. Share 54 by 9. ______
6. 32 divided by 8. ______
7. Digits in 5465454. ______
8. 0·1 × 100 ______
9. 74 × 37
10. 83 × 46
11. 56 × 29

12. **Colour of balloons used**

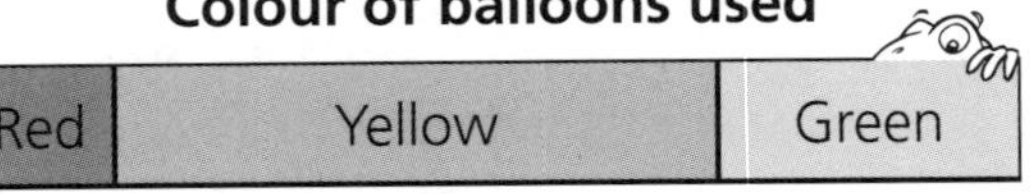

What fraction of the balloons shown in the bar graph was green? ______

13. 9 hectare = ______ m^2
14. a 5 km^2 = ______ m^2
 b 600 ha = ______ km^2
15. The school bought 57 glue pots for $12 a pot. What was the total cost? ______
16. Estimate the size of these angles.

a 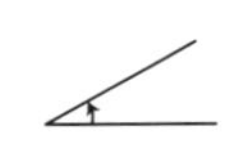 b 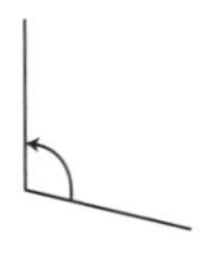 c 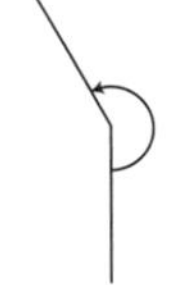

______ ______ ______

17. Colour 6 tenths red and 3 hundredths blue. What fraction have you coloured? ______

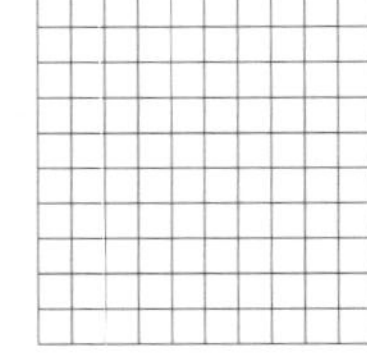

For each pattern, give the number of circles that would be used in the 4th and 10th pictures.

	1st picture	2nd picture	3rd picture	4th	10th
a					
b					

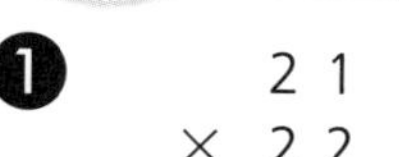

 out of 10

1. 21×22
2. 52×15
3. 61×12

4. Estimate 42 × 12.
E = ______ × ______
= ______

5. Estimate 536 × 34.
E = ______ × ______
= ______

6. This graph shows the response of 24 students. How many chose the category 'bird'? ______

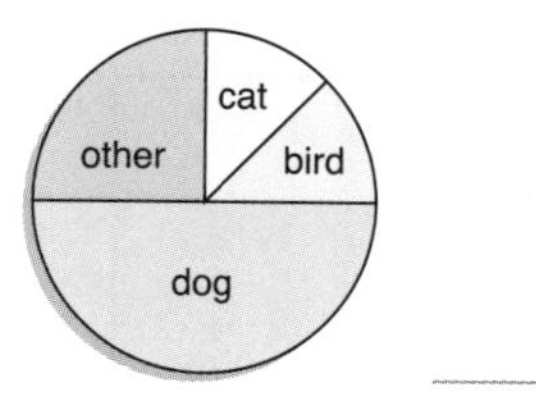

7. 7 hectare = ______ m^2

8. 300 ha = ______ km^2

9.

Games	Piano	TV	Reading

1 cm = 30 minutes

a What kind of graph is this? ______

What fraction of the time was spent:

b watching TV? ______ c reading? ______

How much time was spent on:

d the piano? ______ e games? ______

10.

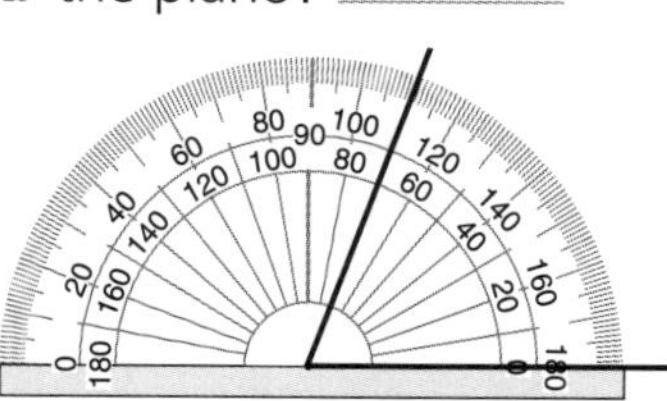

a The size of the angle shown? ______

b Is this angle acute or obtuse? ______

Extension

out of 7

1. Fill in the boxes.

a
$$\begin{array}{r} 3\ 4\ \square \\ +\ \square\ \square\ 1 \\ \hline 8\ 5\ 6 \end{array}$$

b
$$\begin{array}{r} \square\ 1\ 8 \\ -\ 3\ \square\ \square \\ \hline 1\ 7\ 3 \end{array}$$

2. If we buy a 5 kg packet of rice for \$19 every second week for a year, how much will we spend on rice in a year? ______

3. The shortest distance by road from:
a **A** to **F** ______ b **C** to **D** ______

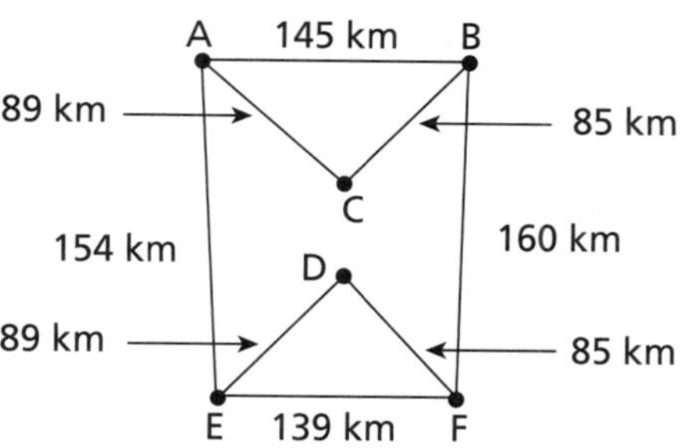

4. a 556 − ______ = 332 b 78 × ______ = 312

5. The Roman numeral for 160. ______

6. Our numeral for CXCIV. ______

7. What is the perimeter of a square that has an area of 25 cm^2? ______

Challenge

List information displayed in this graph.

______
______
______
______

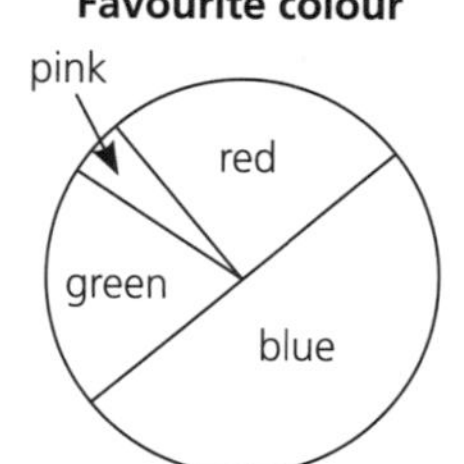

Measure

Fill out this table about yourself, a relative or a friend.

Name: ______ **Date:** ______

Age: ______	Mass: ______ kg	Shoe size: ______
Height: ______ cm	Waist: ______ cm	Neck size: ______ cm

How have these measurements changed since Unit 1 was done?

______

Measurement benchmarks

1
- The **width of your finger** is about **1 cm**.
- The **length of a place-value tens block** is **10 cm**.

2
- The **height of the girl** is a little more than **1 m**.

3
- The **container of milk** holds **2 L**.
- The **can of Fizz** holds **375 mL**.
- The **teaspoon** holds **5 mL**.

4
- The **boy** has a mass of **40 kg**.
- The **margarine** has a mass of **500 g**.

5
- **A double page of the newspaper** has an area of **half of 1 m²**.
- The **top of a place-value ones block** has an area of **1 cm²**.

6
- **30°C** is a **hot** day.
- **3°C** is a very **cold** day.

Use the pictures above to estimate the answers to these questions.

1 a How high is the glass?
b How wide is the table?

2 a How wide would the clothes line be?
b How tall is the mother?

3 a How much will the bucket hold?
b How much will the cup hold?

4 a What is the mass of the dog?
b What is the mass of 2 L of milk?

5 a What is the area of the window?
b What is the area of the top of the matchbox?

6 a What is the temperature on a very hot day?
b What is the temperature on a cool day?

 AUSTRALIAN SIGNPOST MATHS NSW 5 MENTALS • ISBN 978 0 6557 0912 1

Tables of number and measurement

Length

1 centimetre = 10 millimetres
1 metre = 100 centimeters
1 metre = 1000 mm
1 kilometre = 1000 metres

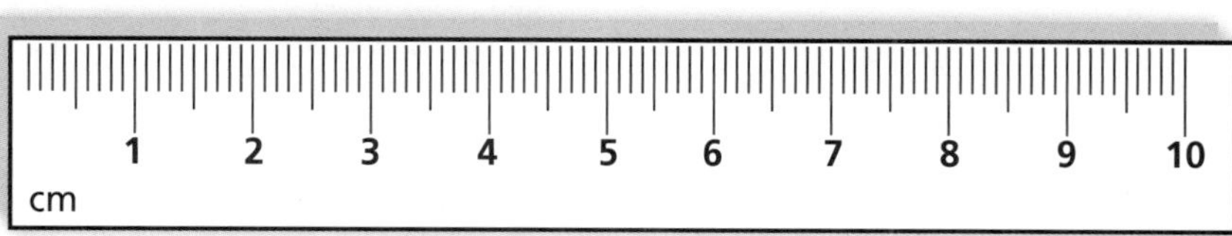

Area

1 hectare (ha) = 10 000 m²
1 square metre = 10 000 cm²
1 square kilometre = 1 000 000 m²
= 100 ha

The **freezing point** of water is **0°C**.
The **boiling point** of water is **100°C**.

°C means degrees Celsius.

A temperature of **5°C** is a **cold** day.
A temperature of **35°C** is a **hot** day.

Mass

1 kilogram = 1000 grams
1 tonne = 1000 kilograms

Capacity and volume

1000 millilitres = 1 litre
1000 litres = 1 kilolitre
1 millilitre = 1 cm³
1 litre = 1000 cm³
1 litre of water has a mass of 1 kg.

A teaspoon holds about 5 mL.

5 mL

A carton of milk holds 1 L.

Roman numerals

1	= I	**6**	= VI	**20**	= XX	**90**	= XC
2	= II	**7**	= VII	**30**	= XXX	**100**	= C
3	= III	**8**	= VIII	**40**	= XL	**200**	= CC
4	= IV	**9**	= IX	**50**	= L	**500**	= D
5	= V	**10**	= X	**60**	= LX	**1000**	= M

Months of the year

Thirty days has September, April, June and November. All the rest have thirty-one, except February alone, which has twenty-eight days clear and twenty-nine days each leap year.

You can use the knuckles of your hands to find the number of days in each month.

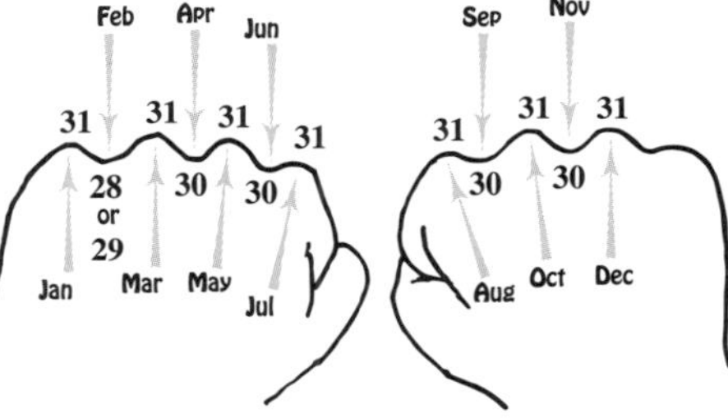

Every 4th year is a leap year.

Time

1 minute = 60 seconds
1 hour = 60 minutes
1 day = 24 hours
1 week = 7 days
1 fortnight = 2 weeks
1 year = 52 weeks
1 year = 365 days
1 leap year = 366 days
1 decade = 10 years
1 century = 100 years

am stands for **ante meridiem**.
am means **before midday**.

pm stands for **post meridiem**.
pm means **after midday**.

Seasons

Summer: December, January, February
Autumn: March, April, May
Winter: June, July, August
Spring: September, October, November

Multiplication tables

1 × 2 = 2	1 × 3 = 3	1 × 4 = 4	1 × 5 = 5	1 × 6 = 6	1 × 7 = 7	1 × 8 = 8	1 × 9 = 9	1 × 10 = 10
2 × 2 = 4	2 × 3 = 6	2 × 4 = 8	2 × 5 = 10	2 × 6 = 12	2 × 7 = 14	2 × 8 = 16	2 × 9 = 18	2 × 10 = 20
3 × 2 = 6	3 × 3 = 9	3 × 4 = 12	3 × 5 = 15	3 × 6 = 18	3 × 7 = 21	3 × 8 = 24	3 × 9 = 27	3 × 10 = 30
4 × 2 = 8	4 × 3 = 12	4 × 4 = 16	4 × 5 = 20	4 × 6 = 24	4 × 7 = 28	4 × 8 = 32	4 × 9 = 36	4 × 10 = 40
5 × 2 = 10	5 × 3 = 15	5 × 4 = 20	5 × 5 = 25	5 × 6 = 30	5 × 7 = 35	5 × 8 = 40	5 × 9 = 45	5 × 10 = 50
6 × 2 = 12	6 × 3 = 18	6 × 4 = 24	6 × 5 = 30	6 × 6 = 36	6 × 7 = 42	6 × 8 = 48	6 × 9 = 54	6 × 10 = 60
7 × 2 = 14	7 × 3 = 21	7 × 4 = 28	7 × 5 = 35	7 × 6 = 42	7 × 7 = 49	7 × 8 = 56	7 × 9 = 63	7 × 10 = 70
8 × 2 = 16	8 × 3 = 24	8 × 4 = 32	8 × 5 = 40	8 × 6 = 48	8 × 7 = 56	8 × 8 = 64	8 × 9 = 72	8 × 10 = 80
9 × 2 = 18	9 × 3 = 27	9 × 4 = 36	9 × 5 = 45	9 × 6 = 54	9 × 7 = 63	9 × 8 = 72	9 × 9 = 81	9 × 10 = 90
10 × 2 = 20	10 × 3 = 30	10 × 4 = 40	10 × 5 = 50	10 × 6 = 60	10 × 7 = 70	10 × 8 = 80	10 × 9 = 90	10 × 10 = 100